SATELLITE-BASED MITIGATION AND ADAPTATION SCENARIOS FOR SEA LEVEL RISE IN THE LOWER NIGER DELTA

Zahrah Naankwat Musa

SATELLITE-BASED MITIGATION AND ADAPTATION SCENARIOS FOR SEA LEVEL RISE IN THE LOWER NIGER DELTA

DISSERTATION

Submitted in fulfilment of the requirements of the
Board for Doctorates of Delft University of Technology
and
of the Academic Board of the IHE Delft
Institute for Water Education
for
the Degree of DOCTOR
to be defended in public on
Friday, 6th April 2018, at 12:30 hours
in Delft, the Netherlands

by

Zahrah Naankwat MUSA
Master of Science in Hydroinformatics
UNESCO-IHE Institute for Water Education
Delft, the Netherlands

Born in Gboko, Nigeria

This dissertation has been approved by the supervisors

Prof.dr.ir. A.E. Mynett   IHE Delft / Delft University of Technology
Dr. I. Popescu   IHE Delft / Politehnica University of Timisoara, Romania

Composition of the doctoral committee:
Chairman   Rector Magnificus, Delft University of Technology
Vice-Chairman   Rector IHE Delft
Prof.dr.ir. A.E. Mynett   IHE Delft / Delft University of Technology, promotor
Dr. I. Popescu   IHE Delft / Politehnica University of Timisoara, Romania, copromotor

Independent members:
Prof.dr.ir. S.G.J. Aarninkhof   Delft University of Technology
Prof.dr.ir. W.G.M. Bastiaanssen   IHE Delft / Delft University of Technology
Prof.dr. F. Martins   Universidade do Algarve Faro, Portugal
Dr.ir. M. van Ledden   World Bank
Prof.dr.ir. J.A. Roelvink   IHE Delft / Delft University of Technology (reserve member)

This research was conducted under the auspices of the Graduate School for Socio-Economic and Natural Sciences of the Environment (SENSE)
CRC Press/Balkema is an imprint of the Taylor & Francis Group, an informal business.

CRC Press/Balkema
PO Box 11320, 2301 EH Leiden, the Netherlands
Pub.NL@taylorandfrancis.com
www.crcpress.com – www.taylorandfrancis.com
ISBN: 978-1-138-60723-1

Dedication

Dedicated to the loving memory of my elder brother and greatest cheerleader, Kwapbial Karel Sallah. You are not here to cheer me bro, but I made it - just like you said I would.

Summary

Accelerated sea level rise (SLR) is the most important climate change impact for coastal areas. The physical properties of deltas and anthropogenic activities make them vulnerable to the effects of the changing climate; however when evaluation of vulnerabilities is needed many coastal deltas lack necessary data for performing such a task. Leaving these coastal areas without adequate plans to combat sea level rise will cost vulnerable areas huge amounts of losses in lives and properties; e.g. as at 1995, it was estimated that in Nigeria alone, a no-response scenario will cost over $18billion in losses including an estimated 17,000km^2 of wetland (Brown, Kebede, & Nicolls, 2011).

Data availability is one of the most important factors for analysis, assessment and modelling of physical and other phenomena related to river and coastal systems. Consequently, to reduce the effects of SLR through adaptation and mitigation, the IPCC recommended that coastal areas collect data on physical, social and economic parameters e.g. topographical, land-use, population, tidal wave and range (Dronkers, et al., 1990). Many developing countries however lack measuring equipment and long-term records; Africa generally lacks long-term observational data to aid hydrological research (Niang, et al., 2014). Long-term shoreline dynamics of the Nigerian coastline for instance cannot be predicted since the available data is insufficient to even explain fluctuations during the last hundred years (Orupabo, 2008).

The Niger delta has the highest sensitivity to climate change in Nigeria (it has a very gentle slope and low elevation) and its adaptive capacity is the second lowest in terms of socio- economic development of the country (FME, 2010). The Niger delta is also one of the coastal areas with little data for coastal planning and management, and consequently has poor availability of data for hydrologic and hydraulic modelling. Quantitative studies on the lower Niger delta have thus been limited by data availability and inaccessibility of parts of the delta (due to insecurity). Consequently, few quantitative studies using in situ data exist on the Niger delta; a quantitative study of the Niger delta by Awosika, et al., (1992) made use of aerial video data to estimate the cost of SLR and erosion losses for the Niger delta. However, subsequent studies by Ericson et al (2006), and Musa, et al., (2016) indicate that the area calculation by the study overestimated the extent of areas to be affected; thus, the value of losses might have been overestimated.

Use of satellite data helps bridge the data gap by providing ancillary data (imagery, elevation, altimetry etc.) that can be used to quantify the effects of SLR on the Niger delta. This thesis uses satellite data and other auxiliary information as the main sources of data for hydrodynamic modelling and GIS analysis. This is a different approach as satellite data in water management and hydrology is normally used as a last resort and not the first point of choice since such data might not have the accuracy and precision of directly measured data. Studies have however shown that innovative methodologies by scientists have enabled better exploitation of satellite data to overcome the limitations and produce results with high correlation and manageable errors within present uncertainties (Musa, Popescu, & Mynett, A review of applications of satellite SAR, optical,altimetry and DEM data for surface water modelling, mapping and parameter estimation, 2015). This study aimed to assess the impact of SLR on the Niger delta land area, coastline, and surface water in an integrated way that will lead to practical recommendations for adaptation.

Using projected global eustatic SLR values of 19mm by 2030 and 35mm by 2050 in addition to subsidence, this thesis estimated that relative SLR (RSLR) for the Niger delta will range from 0.14m–0.44m by 2030, and 0.29m–0.96m by 2050. Using this RSLR values, the results show that a rise in sea levels of 0.14m already inundates areas in the Niger delta, and the flood extent increases with increase in SLR. Consequently, some 4.6–5.2% (viz. 1119.3–1254km^2) of the Niger delta land area can be lost to inundation by 2030, and 4.9–6.8% (viz. 1175.9–1633km^2) by 2050. Furthermore, the results indicate that without subsidence the inundation effect of SLR on the Niger delta will be minimal (since the eustatic values are just 19mm and 35mm by 2030 and 2050 respectively). Subsidence has therefore made the Niger delta very vulnerable to inundation by making the SLR values very high.

Flooding in the lower Niger River will be affected by rise in sea levels especially as the area continues to subside (chapter 4). The effects include earlier occurrence of downstream flooding, increase in water depth and flooding of areas further upstream (than would occur without SLR). This increase in in flooding will be via expansion in lateral flooding extent in the *downstream* areas, but flooded areas will increase *upstream* because higher sea levels downstream will impede downward flow of flood waters which can result in a backwater effect and subsequent flooding of areas upstream. For the years without flooding from upstream, SLR will cause coastal areas downstream of the Niger River to flood earlier than usual. More so areas upstream of the Nun River, which remain dry in normal years, will get flooded when sea levels rise.

SLR will increase the occurrence of coastal flooding (this is indicated by the flood generated by even the lowest level of rise in sea levels) because water levels and water depths will be higher (as shown for example in the Bonny River), thus increasing land area flooding extent. The flow velocity will also increase with SLR, and coastal floodwaters will thus be transported faster along the river to places upstream. Consequently, flooding of land areas at high tide will increase due to higher water levels.

Furthermore, differences in flow velocity around narrow bends will also be higher with SLR than without, making river crossings more dangerous.

A new coastal vulnerability index called coastal vulnerability index due to SLR (CV$_{SLR}$I) was developed in this thesis. The CV$_{SLR}$I evaluates coastline vulnerability due to SLR using 17 physical, social and human influence indicators of exposure, susceptibility and resilience (Chapter 5). The results showed that the variables that make a coastline highly vulnerable to SLR include (i) physical coastal properties, (ii) human influence, (iii) social properties. The reason being that human presence influences variables like coastal infrastructure and high population density, thus increasing the probability of damage to lives and property when a disaster occurs. More so, human interventions on coastal environments can affect sediment supply and accelerate erosion, and should therefore be captured in vulnerability assessments. Besides, the location of many settlements in remote areas, far away from the local government headquarters, reduces resilience to effects of SLR. The combination of these properties make coastal segments highly vulnerable to SLR.

In conclusion, this thesis shows that parts of the Niger delta are highly vulnerable to effects of SLR due to high RSLR, and therefore need adequate mitigation/adaptation measures to protect them. Thus possible coastal mitigation/adaptation interventions for the Niger delta were modelled and studied. This thesis thus recommends that sustainable local resilience practices already being used in parts of the Niger delta should be included in adaptation planning. These include: planting of Bamboo trees for erosion control; using sandbags as bridges and dykes for flood control; using flood receptor pits as temporary flood water storage; and developing community legislation against sand mining and indiscriminate tree felling. In terms of major mitigation/adaptation interventions, measures that can be used for the lower Niger delta include: construction of dykes, by–pass channels, flood–pits (reservoirs), storm surge barriers, coastline shortening and legislation to ensure compliance by all. Furthermore, to effectively adjust to living with SLR in the Niger delta the following strategies should be adopted: building new structures raised on stilts; and changing farm practices to become more resilient.

Samenvatting

Een versnelde toename van zeespiegelstijging is het meest belangrijke effect van klimaatverandering in kustgebieden. De laaggelegen ligging van delta's en de invloed van menselijke activiteit maken deze gebieden kwetsbaar voor de gevolgen van klimaatverandering. Om deze kwetsbaarheid te kunnen kwantificeren zijn meetgegevens nodig, die voor veel deltagebieden ontbreken. Maar als er geen adequate plannen worden gemaakt met maatregelen tegen zeespiegelstijging kan dit verstrekkende gevolgen hebben voor mensenlevens en economische bedrijvigheid. Zo wordt alleen al voor Nigeria geschat dat de schade gemakkelijk kan oplopen tot 18 miljard USD, inclusief een verlies aan kustgebied van ca. 17,000 km^2 (Brown et al., 2011).

De beschikbaarheid van (meet)gegevens is een van de belangrijkste factoren om kustsystemen te kunnen analyseren en modelleren. Vandaar dat het IPCC heeft aanbevolen om met name in deze gebieden gegevens te verzamelen van fysische, sociale en economische processen, waaronder topografie, landgebruik, bevolkingsdichtheid, en getijwaterstanden (Dronkers et al., 1990). Probleem is evenwel dat er met name in ontwikkelingslanden een groot gebrek is aan meetapparatuur en historische gegevens. In grote delen van Afrika ontbreken afdoende waarnemingen om hydrologisch onderzoek te doen (Niang, et al., 2014). Zo kan het lange termijngedrag van de kustlijn van Nigeria niet worden voorspeld aangezien er onvoldoende gegevens beschikbaar zijn om de veranderingen van de laatse honderd jaar te kunnen verklaren (Orupabo, 2008).

De Niger delta is het meest gevoelig voor effecten van klimaatverandering in Nigeria (de delta ligt zeer laag en heeft een heel flauwe helling) en heeft een heel laag adaptief vermogen van socio-economische ontwikkeling in het land (FME, 2010). Er zijn betrekkelijk weinig betrouwbare gegevens om tot planvorming en beheer te komen, laat staan om hydrologische en hydraulische modellen te ontwikkelen. Studies naar de Lower Niger delta worden ook nog bemoeilijkt door de slechte toegankelijkheid van het gebied (vanwege de gevaarlijke situatie). Vandaar dat er weinig gedetailleerde studies bestaan. Awosika et al., (1992) maakte gebruik van luchtfotografie om een schatting te verkrijgen van de schade door zeespiegelstijging en erosie in de Niger delta. Vervolgstudies door Ericson et al., (2006) en Musa et al., (2016) geven aan dat deze schattingen aan de hoge kant waren en dat de schade wellicht overschat was. Door gebruik te maken van satellietwaarnemingen kan het gebrek aan in-situ meetgegevens worden opgevangen en kan toch een schatting worden gemaakt van de effecten van zeespiegelstijging in de Niger delta.

In dit proefschrift worden satellietwaarnemingen gebruikt als de belangrijkste informatiebron voor het ontwikkelen van GIS analyses en voor het hydrodynamisch modelleren van de delta. Dit verschilt van de conventionele aanpak van hydrodynamisch modelleren waarbij satellietwaarnemingen vaak als sluitstuk worden gebruikt en niet als voorkeursbenadering, gezien de beperkte nauwkeurigheid en precisie. Inmiddels zijn er echter innovatieve methoden ontwikkeld die een beter gebruik van satellietwaarnemingen mogelijk maken met een zeer beperkte foutenmarge binnen de huidige meetnauwkeurigheid (Musa et al., 2015). Op basis hiervan is in deze studie een schatting gemaakt van de effecten van zeespiegelstijging en zijn praktische aanbevelingen gegeven voor klimaatadaptatie.

Op basis van de geschatte eustatische waarden van zeespiegelstijging van 19mm in 2030 en 35 mm in 2050, en rekening houdend met voortgaande bodemdaling, komen de schattingen voor relatieve zeespiegelstijging in de Niger delta zoals berekend in dit proefschrift uit op 0.14m–0.44m in 2030, en 0.29m–0.96m in 2050. Berekeningen laten zien dat een stijging van 0.14m al tot overstromingen leidt in bepaalde gebieden van de Niger delta, die steeds groter worden naarmate de stijging toeneemt. Dit heeft als gevolg dat rond 2030 ca. 4.6–5.2% (d.w.z. 1119–1254 km2) van het land in de Niger delta verloren kan gaan door overstroming en rond 2050 zelfs 4.9–6.8% (d.w.z. 1176–1633 km2). De berekeningen laten tevens zien dat zonder bodemdaling het effect van eustatische zeespiegelstijging slechts zeer beperkt zal zijn. Bodemdaling is daarom een belangrijke factor bij het bepalen van de kans op overstromingen in de Niger delta. Andere resultaten van deze studie zijn:

Ook in de beneden rivier van de Niger delta wordt de kans op overstroming belangrijke mate bepaald door zeespiegelstijging in combinatie met bodemdaling. Meer stroomopwaarts wordt de kans op overstroming mede beïnvloed door de stuwkromme die zich naar bovenstrooms uitbreidt omdat de uitstroom in de delta zelf beperkt wordt door zeespiegelstijging. Zelfs als de bovenstroomse rivierafvoer beperkt is, dan nog zullen de laaggelegen gebieden in de Niger delta eerder overstromen dan gewoonlijk, als gevolg van zeespiegelstijging. Datzelfde geldt voor de bovenstroomse gebieden in de Nun rivier, die bij normale afvoeren droog blijven, zullen ook eerder overstromen. Ook voor de Bonny rivier geldt dat zelfs een geringe zeespiegelstijging tot een toename van het overstromingsrisico zal leiden, zoals aangetoond in voorbeeldberekeningen. Ook de stroomsnelheden zullen toenemen, waardoor de afvoer zich sneller langs de rivier zal verplaatsen en tot hogere waterstanden zal leiden. Bij hoge getijdewaterstanden zal dan ook meer land overstromen. Een ander gevolg van hogere stroomsnelheden zal zijn dat het gevaarlijker wordt om de rivier over te steken in de vele nauwe bochten die er voorkomen.

In dit proefschrift is een speciale index ontwikkeld die de kwetsbaarheid van kustgebieden bij zeespiegelstijging kan aangeven: de CVSLRI index. Deze wordt bepaald door 17 indicatoren voor fysische, sociale en menselijke invloeden die bepalend zijn voor de ontvankelijkheid voor en weerstand tegen overstromingen (Hoofdstuk 5). De resultaten geven aan dat de belangrijkste variabelen worden bepaald door (i) fysische eigenschappen van de kust, (ii) menselijke invloeden daarop, (iii) sociale kenmerken. Dit omdat menselijke ingrepen zoals het aanleggen van kustbescherming en bebouwing van de kuststrook, direct van invloed zijn op de kans van falen en het berokkenen van schade in geval van overstromingen. Daar komt bij dat menselijke ingrepen in de kuststrook van grote invloed kunnen zijn op het sediment transport (met name in deltagebieden) en daarmee op de stabiliteit van de kustdelta als geheel. Vandaar dat dit aspect meegenomen moet worden bij het vaststellen van de kwetsbaarheid. In geval van de Niger delta bestaan er veel kleine woongemeenschappen ver weg van lokale overheden, waardoor toezicht slechts beperkt mogelijk is. Ten gevolge hiervan zijn sommige gebieden nog extra kwetsbaar voor de gevolgen van zeespiegelstijging.

Samenvattend geldt dat in deze thesis is aangetoond dat bepaalde delen van de Niger delta bijzonder kwetsbaar zijn voor de gevolgen van zeespiegelstijging en dat adequate maatregelen voor mitigatie en adaptatie nodig zijn om deze te beschermen. Vandaar dat wordt aanbevolen om voort te bouwen op de aanwezige ervaring en bestaande praktijk bij het vaststellen van aanpassingsmaatregelen voor duurzaam kustbeheer in de Niger delta. Daartoe behoren ondermeer: het planten van bamboe struiken ten behoeve van erosiecontrole; het gebruik van zandzakken voor bruggetjes en dijkjes om overstromingen te controleren; het gebruik van kuilen voor het tijdelijk opslaan van overtollig water; het ontwikkelen van lokale voorschriften en verboden tegen het illegaal weghalen van zand en het kappen van bomen en andere vegetatie. Voor de beneden rivier van de Niger in het bijzonder gelden als belangrijkste interventiemaatregelen: het aanleggen van dijken, nevengeulen, opvangreservoirs, het aanleggen van stormvloedkeringen, verkorten van de kustlijn, alsmede wet- en regelgeving die zorgen voor naleving door alle betrokkenen. Om effectieve aanpassing aan het 'leven met zeespiegelstijging' te garanderen zouden de volgende strategieën moeten worden toegepast: nieuwe constructies verhoogd aanleggen, bijv. op palen; landbouwmethoden zodanig aanpassen dat deze meer overstromingsbestendig worden.

Table of contents

1

Introduction

Within the last few decades the atmospheric and sea surface temperatures have been rising and climates worldwide are changing (figure 1.1). Climate change has resulted from an accelerated increase in carbon dioxide and other greenhouse gas concentrations in the atmosphere (Williams & Ismail , 2015). Increase in sea surface temperatures cause thermal expansion, which increase the water level of the sea surface (IPCC, 2013) and as a result the shoreline moves further inland. The warming of the atmosphere causes melting of mountain glaciers and polar ice sheets, thus increasing the rise in sea levels. Based on historical data eustatic sea level changes between 1950 and 2009 were on average 1.7mm/year. In recent years satellite altimetry measurements (between 1993 and 2003) have shown an increase in sea level rise rates to over 3mm/year (IPCC, 2007a). Over the years scientists have used climate models to project possible sea level rise (SLR) levels for the future. These projections are based on scenarios to predict possible conditions of climate change and the states of the coastal areas. In its reports the Inter-governmental Panel on Climate Change (IPCC) had projected a rise of 0.18-0.5 m by the year 2100 (IPCC, 2013). This projection had its limitation due to uncertainties in response of the ice sheets, and their effect on the global sea level. Other projections of higher rise in sea level (e.g. Rahmstorf, 2007; Pfeffer, et al., 2008) were made after the IPCC (2007) report. As data became available, the IPCC revised its projections. Based on greenhouse gas emissions scenarios (known as Representative Concentration Pathways or RCPs), the IPCC projects that sea levels will rise by 0.28-0.98 m by the year 2100 (IPCC, 2013).

Climate change is a factor that will modify existing hazards and introduce new ones (Bogardi, Villagrán, Birkmann, & Renaud, 2005). Natural disasters have become more frequent in coastal areas and barrier islands and river deltas are experiencing accelerated erosion, flooding and marine transgression (Williams & Ismail , 2015). In the Mediterranean region for example, the effects of climate change observed are: decrease in the total amount of precipitation; increase in the number and intensity of extreme events such as floods and droughts; and a change in the seasonal distribution of precipitation (European Environment Agency (EEA), 2012). Along the west African coast, sea levels rise trends showed over 3mm/year from 1993 to 2010 (ESA, Space in Images: mean sea level trends, 2012).

Rise in sea levels has various consequences for low lying coastal areas such as inundation due to coastal flooding by incoming rivers and/or the sea; erosion; displacement of coastal wetlands; and inland intrusion of sea water (IPCC, 2007b; Van, et al., 2012). Furthermore, SLR will reduce the availability of fresh water for human consumption, and affect the fresh water habitat of fishes and other aquatic fauna and flora.

The effects of sea level rise (SLR), however, will not be uniform all over the world but will depend on the physical, socio-economical, and anthropogenic conditions of the coastal area. Consequently some areas will experience higher and more rapidly rising sea levels than others. Relative sea level rise is the change in sea levels relative to the land elevation and includes land vertical movement in addition to global sea level rise values. Relative sea level rise values are therefore higher in subsiding coasts like river deltas than the stable coastal areas.

Coastal delta landforms are formed by the combination of river flow, tides and waves; so that the dominant process determines the characteristic of the particular delta (Nicholls, Wong, Burkett, & Codi, 2007). Fluvial deltas depend on sediment supply from upstream rivers, while Marine dominated deltas are shaped by marine processes of tide and waves. Deltas are usually very rich in biodiversity and are known to expand with increased activity upstream like agriculture and land clearing which loosen the soil adding to amount of sediment transported downstream (Mcmanus, 2002).

Some deltas like the lower Niger delta are also rich in oil and gas (and other mineral resources) making them economically very important to their countries. Coastal deltas are susceptible to subsidence when there is reduction in sediment supply (Wesselink, et al., 2015), and water or hydrocarbon extraction from underground sources (Ericson, Vorosmarty, Dingman, Ward, & Meybeck, 2006). As hydrocarbon/water is extracted, the soil compacts to fill the void and land levels lower; as long as there is normal sediment supply and such extraction is regulated, this process might not be detrimental to the area. However where there is reduction in sediment supply to a delta, the land will subside thereby increasing residence time and reach of high tidal waters the land will subside thereby increasing residence time which can cause water logging and finally permanent inundation (Nicolls, Hoozemans, & Marchand, 1999).

Due to different levels of land subsidence therefore, deltas record different sea level rise values than the global average value; this is known as relative seal level rise. Relative sea level rise values are usually higher in subsiding deltas because it represents the change in sea levels relative to the land elevation and includes land vertical movement in addition to global sea level rise values.

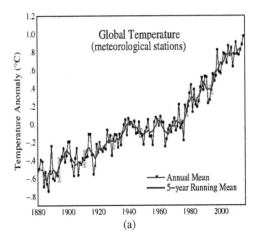
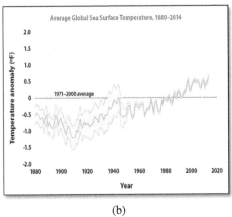

(a)   (b)

Figure 1.1. Atmospheric (a) and Sea Surface (b) temperature rise. Retrieved on 11-10-2016, from
http://data.giss.nasa.gov/gistemp/graphs_v3/, and https://www.epa.gov/climate-indicators/climate-change-
indicators-sea-surface-temperature

In the US, the Gulf of Mexico records a relative sea level rise between 2 - 10mm/year, and the Atlantic coast records between 2- 4mm/year (Titus, et al., 2009). Compared to other coastal areas, river deltas have complex morphologies, because river waters and sediments are transported through the deltas into the sea. A delta can have many elements included, such as barrier islands, multiple estuaries, sand beaches, or mud coasts. It can be crisscrossed by rivers emanating from different sources and carrying different types of sediments; which differentiate the segments of the coast.

Problem statement

In preparation for consequences of SLR, the IPCC recommended that coastal areas collect data on physical, social and economic parameters e.g. topographical, land-use, population, tidal wave and range (Dronkers, et al., 1990). The physical properties of deltas and anthropogenic activities make them vulnerable to the effects of the changing climate; however when evaluation of vulnerabilities is important many coastal deltas lack necessary data (e.g. water level) for performing such a task. Data availability is one of the most important factors for analysis, assessment and modelling of physical and other phenomena related to river and coastal systems. Many developing countries however lack measuring equipment and long term records; Africa generally lacks long-term observational data to aid hydrological research

(Niang, et al., 2014). In well-developed coastal areas such as The Netherlands, measurements and records of hydrologic properties are kept, adequate provision is made for drainage of rain rainwater, river flows and groundwater, and coastal defences are built against storm surges and SLR (Wesselink, et al., 2015).

Leaving coastal areas without adequate plans to combat sea level rise will cost vulnerable areas huge amounts of losses in lives and properties; e.g. in Nigeria alone, a no-response scenario will cost over $18billion in losses, including an estimated $17,000km^2$ of wetland (Brown, Kebede, & Nicolls, 2011). It is therefore imperative to study the level of vulnerability of coastal areas to the effects of SLR like flooding, inundation, erosion, loss of wetlands and salinitization of underground water sources even if data availability is scarce. Sea level data and information are strategic for planning and management of coastal areas, however many developing countries lack measuring equipment and long term records. Long-term shoreline dynamics of the Nigerian coastline for instance cannot be predicted as available data is insufficient to even explain fluctuations in the last hundred years (Orupabo, 2008). The lower Niger delta (figure 1.2) is thus one of the coastal areas with little data for coastal planning and management. Consequently, it has poor availability of data for hydrologic and hydraulic modelling.

The Niger delta has the highest sensitivity to climate change in Nigeria, and its adaptive capacity is the second lowest in terms of socio-economic development of the country (FME, 2010). It is one of the vulnerable coastal areas in the world due to its natural properties: a very low elevation and gentle slope. With an anticipated rise in sea levels of 0.5-1m for the Nigerian coast by 2100 (FME, 2010), large parts of the delta could be affected; with huge costs in both lives and property. Based on physical properties and human population to be displaced, studies by (Brooks, Nicolls, & Hall, 2006) and Ericson, et al., (2006), rank the Niger delta among the vulnerable coasts of the world. Many articles have been written and studies carried out on the possible effects of climate change and sea level rise on the coastal zone of Nigeria (Akinro, Opeyemi, & Ologunaba, 2008; Ogba & Utang, 2007; Awosika, French, Nicolls, & Ibe, 1992); many of these include response strategies and give possible mitigation methods to reduce any negative effects on the land and livelihood of the communities. However a look at the available literature shows that most studies are based on the general Nigerian coast with more emphasis on the Lagos coastal area. Few quantitative studies using insitu data exist on the Niger delta. A quantitative study of the Niger delta by Awosika, et al., (1992) made use of aerial video data to estimate the cost of SLR and erosion losses for the Niger delta. Subsequent studies by

Ericson et al (2006), and Musa, et al., (2016) indicate that the area calculation by the study overestimated the extent of areas to be affected thus the value of loss might have also been over estimated.

Quantitative studies on the lower Niger delta have been limited by data availability and inaccessibility of parts of the delta (due to insecurity). Use of satellite data helps bridge this gap by providing ancillary data (imagery, elevation, altimetry etc.) that can be used to quantify the effects of SLR on the Niger delta. Satellite remote sensing provides a source of hydrological data that is unhindered by geopolitical boundaries, has access to remote/unreachable areas, and provides frequent and reliable data (Jung, et al., 2010). Use of satellite data to estimate hydrological parameters continues to increase due to greater availability of satellite data, improvement in knowledge of and utilization of satellite data, as well as expansion of research topics. A very important catalyst to this growth in satellite data utilization is the ability to use it in a GIS environment. GIS enables comparison and deduction of relationships that exist amongst the complex data sources used for analysis. Thus relationships like the effects of land-use change on surrounding water bodies or water management are easily analysed and depicted. Consequently, satellite data is commonly used for: mapping of water bodies, testing of inundation models, soil moisture measurements, precipitation monitoring, estimation of evapo - transpiration, and mapping of flood extent.

Satellite data have been used for change assessment studies of the Niger delta (Twmasi & Merem, 2006; Ogoro, 2014), flood/erosion extent documentation and mapping (Ehiorobo & Izinyon, 2011; Eyers & Obowu, 2013). Flood inundation estimates for the Niger delta are based on GIS depictions of flooding extends from static overflowing of water on GIS layers; a method that generally over estimates the coverage area of flood waters and does not take loss through local drainage/runoff into account (e.g. Akinro, et al., 2008). This thesis uses satellite data and other ancillary information as the main sources of data for modelling and analysis. This is a different approach as satellite data in water management and hydrology is normally used as a last resort and not the first point of choice since such data might not have the accuracy and precision of directly measured data. However over time, innovative methodologies by scientists have enabled better exploitation of satellite data to overcome the limitations and produce results with high correlation and manageable errors; within present uncertainties (Musa, Popescu, & Mynett, A review of applications of satellite SAR, optical,altimetry and DEM data for surface water modelling, mapping and parameter estimation, 2015).

1.1 Study aims and objectives

The study has the general aim of assessing the impact of SLR on the Niger delta land area, coastline, and surface water in an integrated way that will lead to practical recommendations for adaptation.

The specific objectives will use satellite based data and spatial information to:

- Model the impact of SLR for Niger delta land areas, major rivers and coastlines.

- Create scenarios and run hydrodynamic models of effect of SLR on flooding from surface water.

- Measure and map sea level rise inundation extents based on topography and tidal variability.

- Identify and map the most vulnerable parts of the Niger delta coastline to SLR.

- Create adaptation scenarios for the Niger delta, and estimate the effect of each scenario implementation.

Fulfilling the objectives of the thesis should answer the following questions:

1. How can satellite data be applied in hydrological studies in delta areas?

2. With recent increase in flooding will sea level rise exacerbate the effects of river flooding? What is the effect on surface water?

3. What is the effect of SLR on coastal flooding and inundation?

4. How much of the land that can be lost to inundation?

5. How can the vulnerability of deltaic coastlines to sea level rise be evaluated?

6. What should be considered in planning for SLR adaptation? Are there existing sustainable options that can be used?

7. What are the possible effects of the mitigation/adaptation options on SLR impacts on the Niger delta?

1.2 Scope

The study covers the geomorphologic Niger delta area spanning from Jalla in Ondo on the west to Bonny on the east (Fig. 1.2). The vulnerability assessment is limited to the parts of the Niger delta directly connected to the ocean and inland up to 45km; it does not include the entire political Niger delta as defined by the Nigerian government. The modelling aspect of the methodology only addresses issues concerning river flooding from the Niger River and excludes the following: flooding from other rivers within the area, rainfall/ runoff and other effects of the broad concept of climate change.

1.3 Thesis structure

The thesis is made of eight chapters structured to answer the research questions posed and fulfil the aims and objectives of the research.

Chapter 1: This introductory chapter presents the problem with the study area, and the existing gaps that created the need for the methodology used in the thesis (to solve the problem). The research questions, aims and objectives are also presented herein.

Chapter 2: The second chapter gives the background information on climate change induced sea level rise and its consequences for low-lying delta areas. The study area is described; its physical properties (slope, geomorphology, topography, aquifers etc.), hydrology, as well as the changes it is already undergoing that can be exacerbated by sea level rise. The available data used for the research are also described and their sources mentioned.

Chapter 3: In view of data scarcity, this chapter discusses how advanced data sources are used for surface water and coastal area studies; for modelling, mapping and parameter estimations. Use of high resolution insitu dredging data for river cross section estimation (for flood modelling) is also discussed.

Chapter 4: Using non-conventional data, the effects of sea level rise on river and coastal flooding are modelled, and the results presented in this chapter. The modelling objectives, data used, methodology, model setup, scenarios, and constraints are explained. The implications of sea level rise on flooding extent, water depth, and flood velocity are discussed.

Chapter 5: In this chapter the vulnerability of the Niger delta coastline to sea level rise is calculated based on seventeen physical, social and human influence variables; obtained from both conventional and advanced data sources. The variables are which divided into exposure, susceptibility and resilience variables, are used in a new methodology called the 'Coastal Vulnerability to sea level rise Index' ($CV_{SLR}I$) that combines the 'Coastal Vulnerability Index'

(CVI) with the 'Flood Vulnerability Index' (FVI). The advantages and disadvantages of the new index are discussed.

Figure 1.2. The lower Niger delta Nigeria, West Africa. SRTM DEM showing the topography of the area is

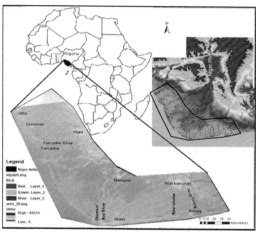

seen on the right.

Chapter 6: This chapter discusses the importance of considering the people's resilience when planning for sea level rise adaptation or mitigation. It considers the sustainability of local resilience practices (against flooding, erosion, etc.) already familiar to the people in a 'business as usual' scenario. A GIS based mapping of sea level rise inundation by 2030 and 2050 is also included.

Chapter 7: In this chapter, adaptation scenarios, possible options they entail, and the implications of each for delta areas are discussed and applied to the Niger delta. The effects on flooding /inundation extent, river water level/depth, and velocity are presented. A rough estimate of the cost of measures applied is also included.

Chapter 8: This chapter compiles the conclusions reached at each stage of the research. It discusses the research questions vis-à-vis the answers obtained within the thesis and makes recommendations where applicable. The limitations of the study and its constraints (based on assumptions made in applying the methodology), are also included.

Appendix: This is made up of two sections A, and B. Appendix A shows how SRTM compares with 40m contour data of Nigeria (using RMSE as accuracy measure). Appendix B shows results of river flood modelling for rivers Forcados and Nun.

2

Background, Study area and Data availability

Generally, deltas are fertile and highly productive, attracting agricultural activities and trade and thus densely populated. For the deltas affected by SLR, the consequences can be devastating; e.g. parts of some coastal areas like the Mississippi delta and Black River marshes in the US have already been submerged by rising sea levels (Titus, et al., 2009). Others like Jakarta (Klijn, et al., 2015) and the Niger delta are undergoing high levels of subsidence and are therefore highly vulnerable to SLR (Syvitski, 2008).

When there is need for coastal protection, the available strategies are water management, sediment management, adjustment of human behaviour, or inaction. Although each of these strategies apply different methods, they can be combined depending on the local characteristics (European commission, 2012). Strategies for mitigation and adaptation can be structural or non-structural; structural measures are used to reduce risk by construction of physical defence mechanisms (e.g. levees, gates, floodwalls, pumps, gates, and weirs), while non-structural measures strategies (e.g. flood warning, change in building regulations/land-use practices, resettlement, buy outs, beach nourishment) that do not change the nature of the hazard event but reduce the consequences by potential damage/loss (US Army Corps of Engineers, 2009).

This chapter is made of three sections. The first presents coastal interventions that have been used around the world to reduce the effects of coastal hazards. The second section introduces the study area, and its characteristics. The available data used for this study are discussed in the third section.

2.1 Adaptation and Mitigation strategies applied on coastal areas around the world

2.1.1 West coast of Africa

The west coast of Africa is undergoing rapid infrastructural development, tourism, fishery, agriculture and urban growth. These population pressure and exploitation of natural resources, have caused coastal environmental degradation (World Bank, 2016). In particular, the areas around the Bight of Benin (figure 2.1), consisting of Ghana, Togo, Benin and Nigeria record significant sand mining and erosion, which lead to loss of land and property. With climate change, coastal flooding has also become more common along the West African coast, with the sea levels showing a steady rise of about 3.6mm/year in recent years (ESA, 2012). SLR with its accompanying effects make it necessary for all coastal protection plans and strategies to be re-evaluated. Coastal protection strategies will now have to stretch beyond protection to include a mix of prevention, mitigation and preparedness (World Bank, 2016).

Studies over the years have shown that the shoreline was quite stable, and affected only naturally by sea levels, however this changed in the 20th century after various human interventions like Damming and Harbour construction (Tilmans, Jakobsen, & LeClerc, 1991). The major coastal intervention projects responsible for serious erosion recorded on this coast are the Lagos harbour (Nigeria, 1908-1912), the Cotonou harbour (Benin, 1960), the Akosombo dam (Ghana, 1963), and the Lome harbour (Togo, 1964). Coastal harbour constructions cause erosion of down-drift areas, thus extending the erosion effects to new areas. For example, after the construction Lome harbour, erosion affected the down drift coastline threatening highly populated areas (Kreme, Agbodrafo, and Aneho) and valuable coastal infrastructure; which necessitated construction of permanent groins to protect those areas. The erosion thus pushed further eastwards, crosses the border to neighbouring Benin republic; however the areas immediately affected are scantily populated and have little economic significance to attract any coastal protection works to be undertaken. Benin republic is instead confronted with the erosion problems caused by the Cotonou port and significant beach sand mining at Seme.

More details are given below on coastal conditions and coastal protection efforts made in two locations on the West African coast in Nigeria and Ghana.

Figure 2.1. Google Earth image showing part of West African coast around the Bight of Benin

2.1.1.1 Lagos (Nigeria)

The 850 km Nigerian coastline consists of some of the most densely populated cities in Africa. The most prominent of them is the city of Lagos with an estimated population of over 20 million and an ever increasing population growth rate (NPC, 2010; World Population Review, 2017). Lagos has a dynamic coastline which has undergone various anthropogenic interventions; from construction of stone moles harbours, to channel dredging, beach sand mining and recent creation of artificial islands. The first coastal interference was the construction of three stone moles (figure 2.2) at the tidal inlet in 1908 which led to almost a kilometre land loss at Lagos Bar beach over the next 100 years (Ibe, 1988). These human interventions have affected the coastline, causing changes in erosion and accretion patterns and locations. These changes include 22-29m/year beach accretion at Lighthouse beach and 20-30m/year erosion at Victoria beach (Ibe, 1988; Orupabo, 2008).

All the morphological changes occurring along the Lagos coast are exacerbated by rising sea levels and land subsidence from lack of sediments and groundwater extraction. The geology of Lagos shows that it is made of Quaternary deposits which are relatively young sediments that are susceptible to compaction (Van Bentum, 2012). Lagos also records frequent storm surges along its coast; in 2006 storm surge caused water levels to rise 1.5m above normal levels, flooding bar beach and surrounding areas. Again, by 2012 Kuramo beach (located to the east) was ravaged by strong waves from storm surge which washed away people and their homes (Ayeyemi, 2013).

To protect the shoreline, several shoreline stabilization measures have been applied around economically viable areas like the Lagos Bar beach. These measures which include: continues sand nourishment (from 1958 – 2006), construction of Groins (one at the foot of East mole parallel to the shore, another east of West Mole), construction of a pumping station on East mole, and construction of a one kilometre long Xbloc revetment; which did not stop the erosion of Bar beach but marginally slowed down shoreline retreat from 2.1m/year to 1.7m/year (Orupabo, 2008).

In recent time a new extension to the city was started on reclaimed land in front of the Bar Beach, behind the East Mole (figure 2.2). This project known as the Eko Atlantic city is protected against erosion with an 8 km barrier at the location of the original shoreline before moles construction in 1908 (figure 2.3). Studies undertaken by (Van Bentum, 2012) show that the Atlantic City project will not reduce overall erosion volumes along the Lagos coast but will shift the location of erosion further east; a need for coastal protection for those locations was thus recommended. Subsequently, in 2013 the Lagos state government announced the commencement of coastal protection projects covering 7 km stretch of shoreline east of the Atlantic City project. The structural protection measures to be used include groins, seawalls, sandbags, floodgates and sand reclamation (Ayeyemi, 2013).

Figure 2.2. Modified Google earth image showing Tidal inlet on Lagos coast Nigeria. Figure above shows the coastline by 2006, figure below shows the coastal modifications made behind the East mole (Atlantic City that extends the bar beach coastline seawards).

Figure 2.3. Eko Atlantic City barrier wall to protect against erosion from waves. It is made of stones piled 15 meters (42ft) high and topped with concrete accropodes. Retrieved from: http://www.ekoatlantic.com/latestnews/press-releases/a-city-rising-from-the-sea-the-dazzling-eko-atlantic-project-in-nigeria/

2.1.1.2 Ghana

The 580 km Ghanaian coast has been severely eroded over the years (figure 2.4), especially in areas around the Volta basin at Keta and Ada (NBCC: National Black Chamber of Commerce, 2011). Ghana has a large fishing population that live along its coasts in fishing villages, these settlements have been severely affected by erosion from natural causes and human intervention through sand mining, marine engineering works, and river basin management (Tilmans, Jakobsen, & LeClerc, 1991). Erosion rates around the Keta area were between 2- 4m per year before sediments from the Volta River were trapped from reaching downstream after the construction of Akosombo dam in 1961; this accelerated erosion rates to 8-10 m/year (Addo, Jayson-Quashigah , & Kufogbe, 2011). Since 1960 when the first sheet pile wall was erected (Nairn, et al., 1998), the shoreline around Keta has had structural coastal defences to protect it from erosion. The largest of such projects is the Keta Sea Defence Project (KSDP, figure 2.5), the first phase of which started in the year 2000 and concluded in 2004. The first phase of the project cost about 90 million dollars to undertake and included dredging works, construction of groins, lagoon flood control structures, land reclamation and resettlement of 1200 households (Baird: Oceans, Lakes and Rivers, 2011; Danquah , Attippoe , & Ankrah , 2014). The KSDP reduced erosion in Keta significantly, but increased erosion rates in the down drift areas east of the KSDP and at areas closer to the Volta estuary (Addo, Jayson-Quashigah , & Kufogbe, 2011).

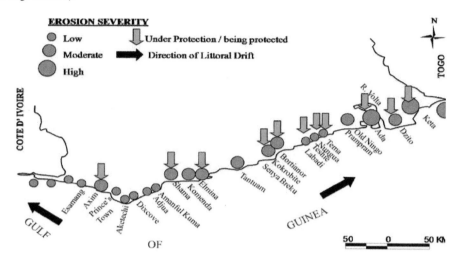

Figure 2.4. Erosion hotspots along the Ghanaian coast (Armah and Amlalo, 1998). Retrieved from https://www.researchgate.net/publication/256756000_Impact_of_sea_defense_structures_on_downdrift_coasts_ The_case_of_Keta_in_Ghana

In 2016, a storm surge hit coastal communities near Keta. Strong waves destroyed buildings and farm lands, deposited plants and refuse, and caused flooding and inundation. 344 households, three schools and over 1720 people were affected. The situation caused the government of Ghana to announce new plans to extend the KSDP to reach those areas in order to protect them from high waves from the ocean (Appiah, 2016).

Figure 2.5. Groin constructed as part of the Keta Sea Defence system on the East coast of Ghana. Retrieved from http://www.imgrum.net/media/978781727019795732_37222592

2.1.2 The Netherlands

About 25% of the Netherlands lies below sea level, and most of the other 75% lie few metres above mean sea level (Schielen, 2010). Due to these unique properties the Netherlands has had to battle with flooding, storm surges, inundation and erosion from the ocean. Since it is also a delta, the Netherlands faces river flooding from the Meuse and Rhine Rivers as they discharge into the North Sea.

Over the years, the people of the Netherlands (the Dutch) have built and developed methods for controlling and mitigating against these hydrological challenges (figure 2.6). The history of Dutch flood protection systems go back almost 200 years with the people digging drainage ditches to keep flood water off their land. By the early 13th century the first 126 km long dykes were built to protect against high water from storms. As effective as the dykes were, they also developed problems like sedimentation which caused raising of the river bed, in-channel sand bar development and ice jams; all of which reduced dyke capacity to hold water and increased possibility of overflow and dyke breaches.

As a solutions to these problems, overflows and lateral diversion channels were constructed as parts of the dykes; these could be lowered to let water out to be redirected to other parts of the river or other rivers. This system however affected parts of land areas through which water had to flow during diversion. To control the flow over land, channel capacities were increased via creation of retention basins and extra flow channels (Oosthoek, 2006).

2.1.2.1 Developments in flood protection

By the 19th century technical improvement made the use of hard structures to control floods more popular, thus the system of flexible flood management was replaced with flood resistance. However, in 1953 coastal flooding inundated a large portion of the land (shown in figure 2.7) and killed over 1800 people (van Arragon Hutten, 2013). The devastating effects of the floods caused the Netherlands government to establish a Delta plan that will protect the country and ensure minimum flood damage in the future. The plan which was made to protect against a 3000 year return period flood reinforced river defences and shortened the coastlines by closing estuaries and inlets with large dams (Oosthoek, 2006; van Arragon Hutten, 2013).

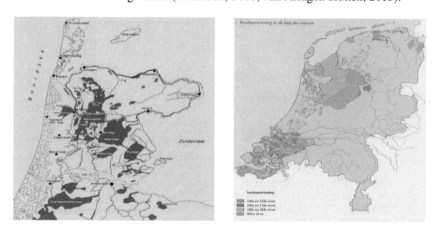

Figure 2.6. (Left) one of the oldest dykes in the Netherlands (red line), built from 1288 - 1300. (Right) coastal interventions showing reclaimed Polder areas from 14th to 20th century. Retrieved from http://home.kpn.nl/keesbolle/Kaarten.html#top

The Delta plan has been revised and modified over the years to include protection against river flooding; this was especially emphasised after the 1993/1995 river high waters/ floods that almost caused dyke breaches in many cities (Utretcht University: Faculty of Geosciences, 2007; Schielen, 2010). Several flood protection measures have been implemented; notable amongst these is the Maeslantkering storm surge barrier made of two huge floating doors that sink when filled with water to protect the city of Rotterdam from North Sea surge (figure 2.8 right).

To commit the government and stakeholders to protecting the country from floods by maintaining pre-determined standards, laws like the 1996 Flood defence Act were enacted. Dyke level standards for example must withstand 1 in 10,000 year storm surge (i.e. a water level of +5m above NAP) around major cities, and those around smaller settlements must withstand 1 in 4000 year storms (Zeeland) and 1 in 2000 (Wadden island) year storms (van der Burgh, 2008; Tulloch, 2016). The Netherlands is protected by 53 dike ring areas with set design flood standards; these are areas protected by dykes, dunes, barriers, dams and high elevated areas (figure 2.12 left). For river flooding, dyke standards are set to withstand 1:1250 design year floods.

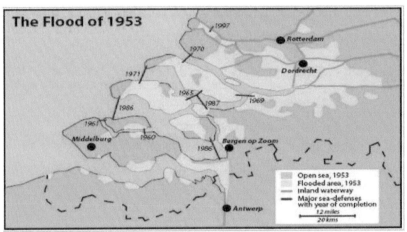

Figure 2.7. The extent of the 1953 flood shown in light blue. The shortened coastline Sea defence systems to protect the western Netherlands and their dates of completion are shown in red (gates, barriers, dams). Retrieved from http://www.mokeham.com/dutchthemag/feature-the-great-flood-of-1953/

Figure 2.8. (Left) showing the dyke ring areas in the Netherlands' Rijnmond-Drechtsteden area. Retrieved from https://link.springer.com/article/10.1007/s11027-015-9675-7. (Right) Maeslantkering storm surge barrier to protect the city of Rotterdam from North Sea surge. Retrieved from http://www.dutchwatersector.com/news-events/news/4311-flood-experts-discuss-need-for-storm-surge-barriers-for-coastal-cities-in-aftermath-of-superstorm-sandy.html

2.1.3 Bangladesh

Bangladesh (figure 2.9) is one of the world's lowest lying countries with only 10% of its land area rising 1m above mean sea level. It is made up of a complex mix of physiographic regions with differences in relief, soil and hydrological patterns. Coastal Bangladesh is divided into three: end of tidal fluctuation zone, salinity intrusion zone, and cyclone risk zone (Karim & Mimura, 2008). Due to the differences in physical characteristics of these zones, effects of SLR will not be uniform across them but will depend on the natural and anthropogenic processes that occur in particular areas. While some areas (e.g. Meghna estuary) are undergoing rapid geomorphological changes, the Ganges Tidal Floodplain is made of stable land (Brammer, 2014).

Coastal intervention measures have been implemented in Bangladesh for several years. The Ganges Tidal floodplain for instance has been protected with embankments (with sluices that allow draining of accumulated rain waters) against tidal flooding since the 50's - 60's. 92 polders were created between 1961 and 1971 covering 10,000 km^2 of intertidal Mangrove areas reclaimed using 4022 km of embankments (Pethick & Orford, 2013). Other interventions like construction of the Farakka Barrage across River Ganges has slowed river flow since 1975, (Brammer, 2014). The different interventions have however increased salinitization of parts of the floodplain and reduced the rate of sediment replenishment for deltaic areas and consequently contributing to land subsidence.

Figure 2.9. Google map of Bangladesh. Hydrological processes are dominated by the combination of river flooding from the three main rivers (Ganges, the Brahmaputra, and the Meghna) and coastal flooding from storm tides and cyclones.

Figure 2.10. An eroded Dyke on the Island of Bhola, Bangladesh. Retrieved from https://www.royalhaskoningdhv.com/nl-nl/nederland/nieuws/nieuwsberichten/20140625pb-nl-experts-ontwikkelen-kustbeschermingsplan-bangladesh/3333

The coastal protection measures across Bangladesh have been inadequate against floods and storms; this is made worse by lack of maintenance of existing structures causing destruction of many dykes (figure 2.7).

2.1.4 Louisiana (USA)

After hurricane Katrina, the authorities modelled 152 hurricanes to enable them design a system to withstand a 100 year strom. Their solutions included a number of structural and non-structural measures to protect against future storm surge risks (that will also be exacerbated by sea levels rise). Hurricane Katrina caused a storm surges up to 8.5 m (and wave run up of 5.5 m) in the Mississippi coast, and 6 m (and wave run up of 2 m) in southeast New Orleans; in addition, it caused levee breaches and floodwall failures (Fritz, et al., 2008). The proposed interventions (figures 2.11/2.12) consisted of multi-scale defence strategies including levees, floodwalls, gated outlets, pumping stations, wetlands restoration, water storage reservoirs, and a surge barrier system (Williams & Ismail , 2015; Woody, 2015).

One of the structural measures put in place is the Inner Harbour Navigation Canal-Lake Borgne Surge Barrier ((IHNC-LBSB), which is a 2.9 km-long surge barrier located at the confluence of the Gulf Intracoastal Waterway (GIWW) and the Mississippi River Gulf Outlet (MRGO). The surge barrier is designed to protect against a 100 year return period event, and provide resilience against a 500 year event. Although the barrier was also designed with a modest 0.3 m SLR factored into the 50 year design lifespan, global eustatic SLR values are projected to exceed the design values (Williams & Ismail , 2015).

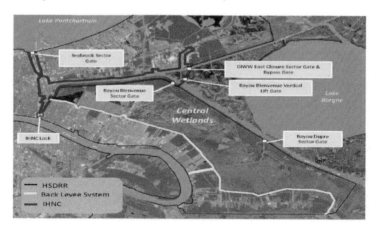

Figure 2.11. Surge/flood protection measures along the Louisiana coast. Gates, levees and surge barrier are shown. Retrieved from (Williams & Ismail , 2015).

For non-structural protection measures, the wetlands restoration plan for the Louisiana coast, seeks to encourage development of a sustainable coastal system by connecting the Mississippi River to the deltaic plain. Sediments brought in by the river will replenish the delta and reduce the land subsidence while enabling wetland expansion and fresh water storage (US Army Corps of Engineers, 2009; Woody, 2015).

Figure 2.12. The IHNC surge barrier system, constructed to hold back storm surge from the Gulf of Mexico.
Retrieved from http://www.takepart.com/feature/2015/08/17/katrina-new-orleans-walled-city

2.1.5   The study area

The Niger delta region (Figure 2.13) is a low lying area consisting of several tributaries of the Niger and other rivers and ending at the edge of the Atlantic Ocean. It has several creeks and estuaries as well as a stagnant mangrove swamp. The region has an area of approximately 20,000km², a 450 km coastline, and is home to about 13 million people. Politically, it is made up of nine states in the southern part of Nigeria where the river Niger breaks into several tributaries. The climate is generally warm and humid with heavy and abundant rainfall average of 2000- 4000mm/year from the rain forest region to the coastal areas. The delta falls within the equatorial rain forest zone, with some parts located in the fresh water swamp and mangrove forest zones. As one goes through the delta in parts of Edo, Ondo, Imo, Abia and Cross River States, however, it is common to see areas that have turned to derived Savanna as a consequence of deforestation and uncontrolled lumbering; such areas are predominantly made of tall grasses, and palm trees.

Nigeria's economy depends on oil and gas extraction from the Niger delta as the main source of foreign exchange, therefore many multinational oil and gas companies operate in the region and over 500 oil wells are located onshore. The activities of oil companies (as well as illegal oil exploiters) like oil spillage and gas flaring have however impacted negatively on the environment, contaminating the surface fresh water and affecting the livelihood and economy of the local people. The Niger Delta is thus faced with many problems which have caused environmental degradation. For example, erosion due to river and coastal flooding has left many areas uninhabited, and acid rain from gas flaring corrodes roofing sheets destroyed biodiversity (Uyigue & Agwo, 2007). The communities around oil companies also experience high temperatures from gas flaring.

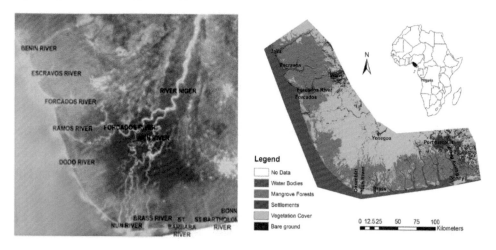

Figure 2.13. (Left) NigeriaSatX image showing rivers in the Niger delta. (Right) The NigeriaSatX image classified into 5 land cover types.

Apart from natural causes, wood logging and deforestation have exposed parts of the soil to more erosion, while onshore oil drilling (from over 500 wells) exposes the land to subsidence and oil spills. Consequently, between 1976 and 1996 there were over four thousand reported cases of oil spills in the Niger delta; Shell petroleum reports 1728 oil spill events between 2007 to May 2015 (Shell N. , 2015).

The economic activity of the Niger delta people is predominantly fishing, farming, lumbering, sand mining and trading. Being a riverine/ coastal area the people are heavily involved in fishery with many fishing communities located at choice areas (Bachmair, et al., 2012). Though there are many oil companies in the Niger delta, only a small number of the local people are employed there, more so, government jobs are few and cannot absorb the growing population. Consequently, the unemployment rate is one of the highest in the country (NDRMP, 2004a;

FME, 2010). Between 1991 and 2006 the population of the Niger delta increased with a growth rate of over 3.1% (NPC, 2010). The growth rate was partly due to the creation of more states in the region and the consequent increase in trade/commercial activities. State creation brings government ministries and agencies as well as non-indigene settlers and visitors; encouraging building of hotels and expansion of residential areas. Consequently rapid changes in land-use/ land cover occurred within the Niger delta (Shell E. , 2004). With such increase in population, more people are exposed to hazards in the delta. Large populations increase the value of risk an area is exposed to because human settlements come with infrastructure, farming, and other economic activities which can be affected by hazards.

Based on population figures of 1995, Awosika, et al., (1992) estimated that 600,000 villagers in the Niger-delta would be displaced in case of a 1m sea level rise. This estimate however may be surpassed as the population of the Niger delta increased with a growth rate of over 3.1% between 1991 and 2006 (NPC, 2010). Many of the problems in the Niger delta will be exacerbated by SLR (Ericson, Vorosmarty, Dingman, Ward, & Meybeck, 2006; IPCC, 2007b; NEST, 2011; Musa, Popescu, & Mynett, 2014a). Besides, the Niger delta has already recorded tidal water levels in areas that had hitherto not been reached by coastal waters (NDRMP, 2004a).

2.1.5.1 Hydrology

The Niger delta has abundant water sources including: creeks, lakes, estuaries and several streams (figure 2.14a). Apart from the River Niger which supplies most of the river discharge into the area, there are many other rivers that drain into the Atlantic ocean throughout the delta; some of these rivers are very large e.g. Qua Ibo, Imo and Cross river, others are smaller e.g. Orashi and New Calabar. These rivers are independent of the Niger River, originating from upland areas with some as far as the Cameroun Mountains.

The Niger River bifurcates into the Nun and Forcados rivers as it flows through the Niger delta, with the Forcados Rivers taking 46% of the discharge and the Nun River taking 54% (NDRMP, 2004a). The Imo River which is the second largest after the Niger is located to the east of the River Niger; it flows through 107km southwards and has a width range of 60m at the upper reaches and over 1000m at the estuaries. All the rivers drain into the Atlantic Ocean giving the delta its complex structure of land and estuaries (figure 2.14a). The Nigerian coast records higher sea levels between September and October (Nwilo, 1997); this coincides with the rainy season upstream and flooding periods in the Niger delta (NDRMP, 2004a). Figure 2.14 (b) shows measured flow and river levels upstream of the delta; with low flow periods from December to April, a high flow period from April to October when there is peak rainfall from the north and locally, and a gradually declining flow period between October and December (NDRMP, 2004a).

According to the Niger delta Environmental and Hydrology report (NDRMP, 2004a), the Niger delta is made of two zones, the coastal zone and the fresh water zone. These two zones undergo different types of flooding as summarized below:

"The coastal zone, dominated by tidal activities, extends inland to about 50km in some places and comprises largely mangrove swamps and beach ridges. This zone is subjected to diurnal inundations, strong tidal currents, waves and floods, especially during high tides. The fresh water zone on the other hand is inundated annually and exposed to strong/hydraulic currents, which erode and cause the rivers to modify and sometimes abandon their course" (NDRMP, 2004a).

In recent times, the coastal zone has experienced peak floods between September and October every year, with the lowest lying communities having diurnal floods. Floods in 1988 and 1994 displaced hundreds of communities in the Niger delta causing loss of farms, homes and sacred sites; the inundation from the 1988 flood lasted for two weeks. In 2012, Nigeria experienced the highest flood ever recorded (Figure 2.14c), when the Niger River experienced a 10-18 year return period flood from August to September; the consequence of this flood is well documented and shows extensive damage in the Niger delta states of Delta, River, Bayelsa, and Edo (NEMA,Nigeria, 2013).

The Niger delta also experiences flash flooding from the high rainfall intensities per hour which can be as high as 135mm/hr over a 15 minute period or 310mm/day, in places like Bonny, Calabar and Warri (NDRMP, 2004a). More so, the Nigerian coast records storm surges between April and August due to swells caused by low pressure in the Southern Atlantic Ocean. During such times tide records in the Niger delta can be high; for example, tide measurements at Bonny recorded water levels up to 50cm above high tide levels (2.9m spring tide) during a storm surge (NDRMP, 2004a). As a consequence of flooding, the Nigerian government at the state and federal levels have over the years applied some structural flood and erosion control measures along rivers in the area like: sand nourishment, revetment, sheet piling, Reno mattress, earth dykes, and concrete piling. Reports by the United Nations Framework on

Convention on Climate Change (UNFCCC) indicate that the Niger-delta region could be inundated with water due to the effects of climate change; and it has the highest risk of water inundation in the gulf of guinea according to Brooks, et al., (2006). The Niger delta is also said to be one of the storm zones likely to be affected by increase in frequency of storm surges in the world according to a study by Dasgupta, (2009) with urban areas like Baguma and Okrika falling within the top ten cities to be affected.

Unfortunately, the Niger delta lacks effective gauging stations where hydrological measurements can be taken, to enable effective mitigation and adaptation strategies to be implemented. It is extremely important to create a database of factors related to SLR like: data on rivers (water levels, flows, bathymetry, water quality), land (topography, land-use, land-cover), coastal estuaries (water quality, water levels/flows) wetlands (area, biodiversity, vegetation), near shore (area, water levels/ tidal changes, land-use etc.).

2.1.5.2 Aquifers

The type of aquifer in an area and the geological formation of the aquifer bedrock determine the ease of salt intrusion into underground aquifers. Salt water intrusion into both surface and ground water aquifers is another effect of SLR. The geomorphology of Nigeria divides the coast into four zones: the barrier lagoon coast, the mahin mud coast, the Niger delta and the Strand coast (Awosika & Folorunsho, Nigeria, 2012). The Niger delta is underlain by eight different geological formations: the Crystalline Basement Complex, Cross River Group, Ajali Sandstone Formation, Nsukka Formation, Bende-Ameki Formation, Benin Formation, Deltaic Plains Formation and the Alluvial Deposits. The regional aquifers within these formations have different properties and suitability for domestic uses (NDRMP, 2004a). The areas closest to the coast in the Niger delta are underlain by Alluvium, Deltaic Plains and Benin Formations which contain water with high salinity. The people living in those areas therefore use surface water and rain water as the main sources of drinking water (since the groundwater is not drinkable). For areas above the immediate coast, there are settlements that depend on borehole water as the primary source of drinking water; such areas will be vulnerable to contamination as a result of intrusion of sea salts.

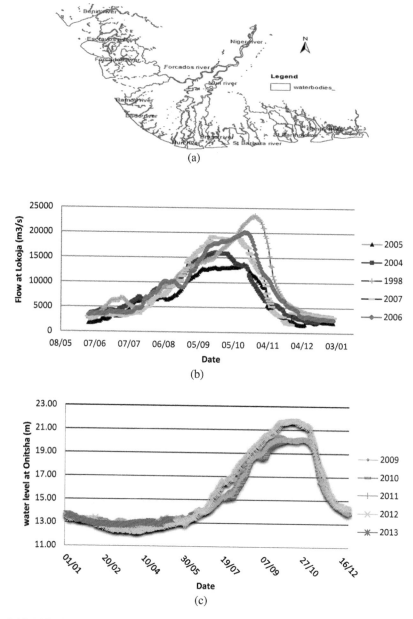

Figure 2.14. (a) Drainage of the lower Niger delta, and names of some major rivers. (b) Flow hydrographs for Niger River at Lokoja. (c) Water level hydrograph for Niger River at Onitsha

As global sea levels rise, estuaries will be affected by salt water migrating upstream. Sea water intrusion will affect the natural biodiversity of estuarine plant and animal species and where the estuary recharges an aquifer, the water will turn salty (Sorensen, Weisman, & Lennon, 1984).

Brackish water vegetation such as mangroves are very sensitive to levels of salinity in their environment and will be affected by too much salinity if the area is covered by salt water; the effects of SLR will however be positive in hyper-saline coastal areas like lagoons as the increased tidal action will reduce salinity and increase water renewal (Nicholls, Wong, Burkett, & Codi, 2007).

2.1.5.3 Shoreline

The Niger delta shoreline is made up of sandy barrier islands, interrupted by estuaries. The barrier islands are generally flat with maximum elevations just above the Highest Astronomical Tide level (HAT); between HAT + 0.2 meters to HAT + 0.5 meters (NDRMP, 2004a). The type of coastal shoreline in an area determines the amount of erosion that will affect it; beaches will experience an accelerated rate especially in areas that are already eroding, estuaries will experience less erosion than inundation and land loss, reefed coasts will have higher wave action and increased water depth that will cause erosion (Sorensen, Weisman, & Lennon, 1984). Coastal areas undergo erosion due to movement of tides and waves in and out; erosion removes soil from land and degrades its ability to protect the coast and support the ecosystem. Although shorelines often change in response to oceanographic perturbations like storms surges, wave energy and sediment supply changes, they always adjust to maintain a morpho-dynamic equilibrium state (Nicolls et al., 2007).

The Niger delta shoreline has been undergoing erosion, as shown by the Nigerian Institute of Oceanographic and Marine Research (NIOMR). Using beach profiling, NIOMR measured the yearly rate of beach erosion along the Nigerian coast and the result showed a range from 14m/year to 25m/year. The erosion ravaging the Niger delta is due mainly to natural causes like river flow and ocean surge, and also by construction of bridges, canals and other coastal structures which altered the natural course of the rivers; thus the area records gully, bank, ravine, sheet, and coastal erosion. The most disturbing trend is the loss of natural flood protection measures like mangroves; erosion has also led to loss of oil wells in the Niger delta (Uyigue, Climate change in the Niger delta, 2007). For changes due to sea level rise, shoreline equilibrium state can be maintained if the rise in sea levels is slow over time and sediment supply is maintained, however if SLR is accelerated, the rate of local coastal erosion will increase beyond the threshold and shoreline equilibrium will not be attained. Rising sea levels will exacerbate beach erosion, undercut cliffs and rocky coasts and cause instability of estuarine systems; however the response will depend on the sediment budget of the coastal area (IPCC, 2007a).

Anthropogenic activities, as well as terrestrial processes like rainfall/runoff all add to the complexity of coastal processes, and have to be considered when predicting effects of SLR on beach erosion. Until recent years, the effect of SLR on coastal erosion was calculated using Bruuns rule as 100 times the rise in sea levels. This method concluded that erosion will be less devastating effect of SLR on coastal areas than inundation (Brooks, Nicolls, & Hall, 2006). However a recent model which includes anthropogenic and terrestrial process in predicting effects of SLR on coastal erosion has shown that results based on Bruuns rule underestimate the effect of coastal erosion by 25-50% (Ranasinghe, Duong,, Uhlenbrook, Roelvink, & Stive, 2012) .

2.1.5.4 Wetlands

The Niger delta has a low lying mangrove swamp that extends up to 50km inland and rises up to 2m above MSL. The Mangrove swamp is located at a transition zone from the fresh water zone in some parts; where Mangroves grow on mud banks of sediments deposits from rivers. The soil is mostly silty and easily saturated which reduces infiltration of rain water; this coupled with a very low gradient induces regular urban floods (NDRMP, 2004a).

Coastal wetlands are located on transition zones between land and sea and are therefore adapted to changing levels of salinity, and changing water levels due to tidal effects. Wetlands are highly productive, and host important ecosystems like coral reefs, sea grass beds, mangroves, and estuaries. In the tropics, wetlands are usually forested and vegetated with e.g. mangroves, while in temperate regions they are grasslands known as salt marshes. Although wetlands are home to different bird and fish species, and host a variety of biodiversity they are little appreciated and are in danger of being destroyed (Bacon, 1996). The location of coastal wetlands are intimately linked to sea level and therefore long-term changes in sea levels will affect them; up to 22% of wetlands might be lost globally due to effects of sea level rise (Nicolls, Hoozemans, & Marchand, 1999). Depending on the surrounding land-use, wetlands also adapt to changes in sea level by migrating; natural features like mountains/ plains can allow adaption, thus the wetlands can migrate inland/ upwards, while man-made features like roads/ settlements etc. are not adjustable and will therefore impede the natural processes that allow adaptation (Nicolls, Hoozemans, & Marchand, 1999; CIESIN., 2013). Flooding due to sea level rise (SLR) can cause permanent inundation of wetlands when the flood waters are impeded from flowing into the sea by high sea levels. The Niger delta Mangrove swamp is inundated daily and might be lost at a cost of over $8m/year in the event of SLR (Brown, Kebede, & Nicolls, 2011).

2.2 Coastal protection for the Niger delta

This thesis aims to propose adaptation/mitigation measures for sea level rise for the Niger delta. From the examples of coastal protection measures discussed in section 2.1.1 above, proposed measures will depend on the characteristics of the study area. Since the lower Niger delta like Bangladesh and the Netherlands is influenced by both river and coastal flow, this study will follow the examples of both countries.

With current climate change effects (of hotter temperatures, higher precipitation, SLR) in addition to land subsidence, the Netherlands now has a Delta Program that combines protection from coastal and river flooding; it uses a strategic risk-based approach to flood risk management that also ensures availability of fresh water (Government of the Netherlands, 2008; Nillesen & Kok, 2015). In the risk-based Delta program, flood standards are set so that human loss of life is the most important criteria and the standard is to protect the country to a life protection level of 1:100,000 years of becoming a flood casualty for all living close to a dyke/dune (Local individual risk, LIR). To implement the Delta program the Netherlands has gone back to flexible water management. Thus apart from construction of dykes/dunes/ barriers/ sluices and installation of pumps, the Delta program includes resilient cities planning, dune sand nourishment, and implementation of a 'Room for the River program'. The Room for the river program which reduces the probability of river flooding by dyke reinforcement, removing floodplain obstacles, river bed deepening, excavating the entire floodplain, enlarging the floodplain area by relocating dykes, constructing spillways/overflow channels and retention basins (Nillesen & Kok, 2015).

Studies of ongoing projects in Bangladesh show that to protect the coast against tidal activity and storm surges, as well as prepare for SLR, a contract for construction of coastal defences was awarded (Cardno, 2016). In the project plan, structural measures will be used to protect the polders, upgrade 618 existing embankments, sluices and channels to flood-safety requirements. The sluicing gates which were previously incorporated into the embankments to let water will now consist of sliding gates as part of an agricultural plan that will also let in water during droughts. Non-structural measures includes planting of Mangrove and other salt-Tolerant species to reinforce the embankments with vegetation. The contract known as Coastal Embankment Improvement Project, Phase-1 (CEIP-1) closely resembles the Netherlands Delta Program (Cardno, 2016).

Similarly, following the Netherlands example this thesis will consider different strategies for SLR impacts reduction, it will determine most vulnerable areas, the type and extent of vulnerability, the types of interventions that might effectively protect an area, and generate the data needed to model and quantify the interventions. At the end, the best intervention method (if any) will be proposed for the Niger delta.

2.3 Available Data

Assorted literature on the Niger delta were read and mined for any information that will be useful as a source of data for this study. Examples of such literature include Shell Nigeria EIA reports, World Bank sponsored study reports, FIG publications, Amnesty international reports, etc. These reports have information about various projects in the Niger delta and usually contain information about the hydrology/environment and can provide some measured data.

Apart from literautre, data available for this research came from several sources: National Space Research and Development Agency (NASRDA), Nigerian Hydrological Services Agency (NIHSA), Nigerian Port Authority (NPA), Nigerian Inland waterways Authority (NIWA), websites of the following government and non-governmental bodies: National Population Commission (NPC), Nigerian Institute for Oceanographic and Marine Research (NIOMR), National Emergency Management Agency (NEMA), Niger Delta Development Commission (NNDC), World Bank, Google earth, Dartmouth flood observatory, Shell Nigeria limited, and USGS-Eros data centre. More details about the different types of data are explained in the following sub-sections.

2.3.1 Measured discharge data for Lokoja upstream of the Niger delta

These are excel table data obtained from the Nigerian Hydrological Services Agency (NIHSA), consisting of twenty eight years of flow discharge measurements (from 1980 – 2008) for Lokoja gauging station. This data, some of which are shown in figure 2.14 (b), had enough information to show the trend of flow at Lokoja from 1980 – 1992, and 1999 – 2008; the data from 1993- 1998 contained many missing measurements.

2.3.2 SRTM DEM

Shuttle Radar Topography Mission (SRTM), was a combined mission of the National Imagery and Mapping Agency (NIMA) and the National Aeronautics and Space Administration (NASA) to generate digital topographic data for 80% of the Earth's land surface. SRTM DEM data at 90mx90m and 30mx30m resolution were downloaded in different tiles of 5deg x 5deg (Geotiff format) from USGS website (earthexplorer.usgs.com). The DEM tiles were mosaicked to cover the lower Niger River basin (figure 2.15).

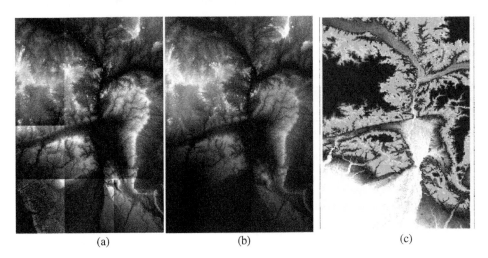

(a)   (b)   (c)

Figure 2.15. SRTM DEM of the lower Niger basin. (a) Tiles before mosaicking, (b) and (c) after mosaicking.

2.3.3 2007 Flood map

In 2007, a flooding event occurred in the Niger River basin. By 26th of July many tributaries of the Niger River overflowed their banks in Burkina Faso, Mali, Niger, Nigeria, Ivory Coast, Ghana, Togo, Benin, Sierra Leone, Liberia, and Guinea. The flood, which affected villages, towns and farmlands lasted until 21st of October in the lower Niger basin (Brakenridge, Kettner, Slayback, & Policelli, 2007). With inadequate measuring, monitoring and mapping equipment in the lower Niger basin in Nigeria, only satellite data from the US MODIS satellite captured the flood as it occurred daily. The flood map (figure 2.16) created based on the MODIS satellite data is available from http://www.dartmouth.edu/~floods/2007144.html.

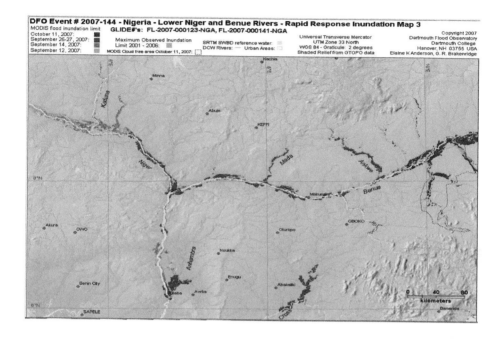

Figure 2.16. Map of flooding in the lower Niger basin, Nigeria.

2.3.4 Shape files of: contours, state boundaries, Local Government areas, towns and settlements, rivers and streams (Nigeria, built up areas (2008).

These GIS based data were obtained from the National Space Research and Development Agency (NASRDA). Data on towns, settlements and local government areas were obtained by the agency during one of its data collection projects in collaboration with the office of the surveyor general of Nigeria. The data on river, streams, contours and boundaries were digitized from topographic maps of Nigeria obtained from the survey agency. Projects that gather data on a national level are rare and far between; thus any updates of these data if available, will cover limited areas of interest. Figure 2.17 shows a 40m contour map of Nigeria created from the contour data; the Niger delta with elevations lower than 40m is completed cut off.

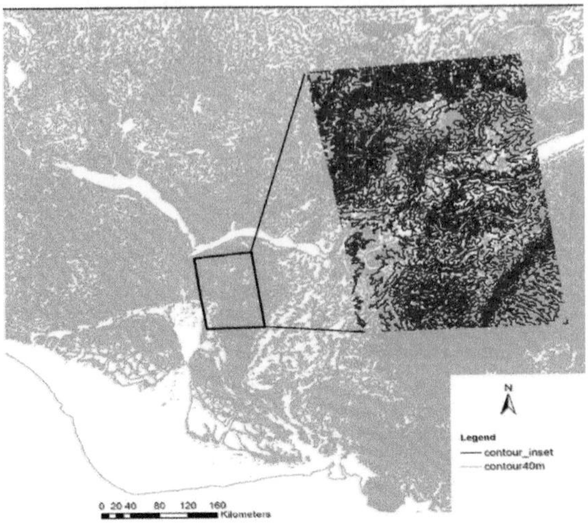

Figure 2.17. 40m contour map of Nigeria; purple inset shows magnified contours of the boxed area.

2.3.5 Satellite imagery

Satellite imagery covering the study area were obtained from NASRDA. The imagery include NigeriaSat1 image from 2005, and NigeriaSatX image from 2013. The freely available LandSat7 and 8 imagery were also downloaded from the United States Geological Surveys (USGS) Eros data centre website. NigeriaSat1 image has 32m spatial resolution and has three spectral bands: near infrared, red and green; NigeriaSat X also has same spectral bands but has 22m spatial resolution. Landsat 7 and 8 have 30m resolution except the panchromatic and thermal bands which have 15m and 100m resolutions respectively. NigeriaSat1 and NigeriaSatX images are shown in figures 1.1, and 3.1.

2.3.6 Dredging data on the Niger River

5-10m resolution survey water depth, and gauge water level data (figure 2.18) were obtained from Nigerian Inland Waterways Authority (NIWA). The data collection was done from December 2001 to January 2002 during a survey operation in the Niger River in preparation for river dredging. The survey used two digital echo sounders to measure water depth, and Real Time Kinematic (RTK) positioning equipment in GPS mode, calibrated using a coordinated benchmark station offshore. The fluctuation in GPS Eastings and Northings were less than 5m, which is acceptable for a thalweg /channel crossing survey (El-Rabbany, 2002).

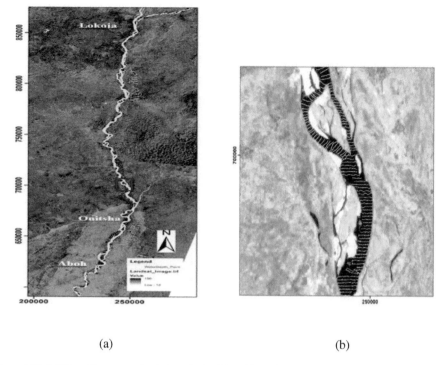

(a)   (b)

Figure 2.18. (a) Niger River water depth data obtained from December 2001 to January 2002. The area within red box is shown in (b).

2.3.7   Topographic data for parts of Eastern Niger delta

GPS readings in XYZ format were obtained from the University of Port-Harcourt River state in the Niger delta. The GPS data were taken for a particular project for the state, and are therefore limited to the projects areas of interest namely: Bonny, Port-Harcourt, Ituk, Degema, Penniton, Oliobiri and Kula. Although the data measurement is concentrated along roads, and the distribution is scanty (figure 2.19), however, since this is a data scarce area, it still provides some ground-truth data to compare with SRTM DEM data.

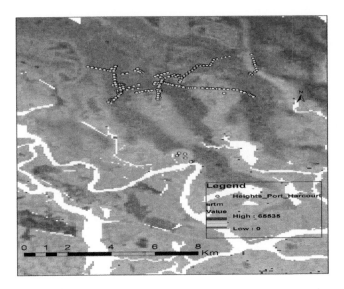

Figure 2.19. GPS spot heights of parts of Port-Harcourt over laid on SRTM data. The data coverage is limited to the areas of interest of the original project for which it was obtained.

2.3.8 Niger Delta Regional Master Plan, Environment and Hydrology report.

This report, prepared as part of a development master plan for the Niger delta covers important thematic areas like the quality and characteristics of surface/groundwater, erosion, socio-economy, flooding, etc. The report which was prepared and submitted is 2004, is quite comprehensive and provides information that can be used as benchmarks for conditions in the Niger delta as at 2004; it also contains recommendations for future planning, which are very useful for projects such as this one. One part of the report was downloaded from the NDRMP website, while other parts were obtained via a personal networking effort. The complete report consists of PDF copies, spreadsheets, and GIS/ CAD datasets.

2.3.9 Socio-economic data (population, water supply/demand, available water resources)

Data on the last Nigerian population census held in the 2006 were obtained from the National Population Commission (NPC) website and via NASRDA. The data from the website consists of PDF documents with the breakdown and analysis of the census data based on states, local governments (LG), age, and gender. The information also contained comparisons between the last two census held in Nigeria (1991 and 2006). The data from NASRDA consists of excel spreadsheets containing the census data per state and LG areas.

Data on water supply and demand in the Niger delta were obtained from the Niger delta regional master plan (NDRMP) environment and hydrology report.

3

Extracting information from modern data sources[1]

---

[1] This chapter is an edited copy of the journal publication: Musa, Z. N., Popescu, I. & Mynett, A., 2015. A review of applications of satellite SAR, optical, altimetry and DEM data for surface water modelling, mapping and parameter estimation. Hydrology and Earth Systems Sciences (HESS), 19, 3755-3769

Due to non-availability of measurement devices, inaccessibility of some terrains and limitations of space/time, hydrological data collection still remains a difficult task nowadays (Quinn, Hewett, C.J.M., Muste, M., & Popescu, I, 2010; Pereira-Cardenal, et al., 2011). In developed nations, efficient infrastructure of flood monitoring and management results in a wide range of high-cost modern data and methods such as high resolution terrain data and digital elevation models, cloud penetrating radar data which defines flood extents (Sanyal & Carbonneau, 2012). Many countries around the world have however recorded a decrease in deployment of physical infrastructure for hydrological measurements. In developing countries data is hardly available, and where it is available, it is sparsely distributed so that data provided is insufficient and limited in scope (Musa, Popescu, & Mynett, 2015; Price, 2009).

Therefore to study ungauged river basins in data scarce areas, several methods have been used by researchers; including use of alternative data sources like satellite data, data interpolation from areas with similar physical characteristics or within same geographical location, and synthetic data development using mathematical/statistical methods (Price, 2009; Sugiura, et al., 2013; Kiptala, Mul, Mohamed , & Van der Zaag, 2014; Musa, Popescu, & Mynett, 2015).

This chapter presents applications of two alternative data sources for hydrological research. The first part is a review of satellite data application in surface water modelling, mapping and parameter estimation. The second part presents an example application of high resolution dredging data for flood modelling.

Use of satellite data sources

3.1 Introduction

A good alternative to overcome data scarcity is use of satellite remote sensing, which can give a synoptic view of target areas (figure 3.1), measure target surface changes and therefore provide information needed for hydrological studies, river basin management, water hazard/ disaster monitoring/prevention and water management, etc. Through the science of remote sensing, information about an object can be obtained without coming in direct contact with it (Lillesand, Kiefer, & Chapman, 2004). This capability works by measuring electromagnetic energy reflected or radiated from objects on the earth's surface (figure 3.1), in such a way that the difference in reflectivity of objects enables recognition/detection and isolation of each type/class.

Remotely sensed data are of two types depending on the main source of energy. Passive remote sensing depends on natural energy from the sun. Active remote sensing uses controlled energy sources from instruments beaming sections of the electromagnetic spectrum. Imagery obtained via instruments that measure reflectance from the sun, are known as optical imagery. Optical imagery from satellites is therefore acquired during the day since it depends on the reflections of sunlight from objects on the earth surface in the absence of cloud cover. Depending on the mission specifications satellites are placed on different kinds of orbits around the earth. The orbits include: Low Earth Orbit (LEO), Medium Earth Orbit (MEO), and Geo-Synchronous orbits (GSO); variations of these classes of orbits are the polar orbit, the Geostationary orbits, the Molneya orbit and the sun-synchronous orbit. Most optical satellites used for hydrological applications are in near earth orbits and are therefore able to provide detailed data at high ground (e.g. figure 3.1); although the best resolution data are usually not freely available and expensive to obtain. Due to this detailed resolution, optical satellite imagery is used for inundation mapping, drainage mapping, disaster monitoring, land-use/land cover change analysis etc. (Owe, Brubaker, Ritchie, & Albert, 2001).

Active remote sensing can provide data as imagery (e.g. radar), and in the form of pulse measurements (e.g. altimeters and scatterometers). Radar is an active source of remote sensing data which acquires data via instruments that emit radar signal towards the object of interest and measure the reflected energy from the object. Radar can penetrate cloud cover and can be acquired at any time independent of availability of sunlight. The penetration characteristic of the SAR satellites enables measurement of soil moisture in bare areas, making it useful for land-use and land cover studies as well as earth observation and monitoring (Owe, et al., 2001). SAR is a side-looking instrument that sends out signals inclined at an angle. For water bodies the reflectivity of SAR waves is spectacular giving a very low radar return and very dark images. However when there are surrounding or emergent vegetation, wind, turbulence etc., there can be significant backscatter; which affects the accuracy of information obtained from the radar measurements (Smith L. C., 1997).

Satellite remote sensing has been applied in hydrology for many years. Table 1 shows some satellite missions and sensors used for hydrological studies and the application areas. A review by Smith, (1997) shows that the earliest hydrological applications were in water body and flood mapping; the review includes many examples of inundation maps developed from satellite imagery. Owe, et al., (2001) also compiled papers presented at a conference on applications of remote sensing in many aspects of hydrology. Beyond mapping, satellite data in the form of imagery, DEM, altimetry data, etc., can be used as hydraulic model input forcing factors or to constrain model data during calibration/validation/verification (Pereira-Cardenal, et al., 2011). Satellite based estimates of river flow, river width, water levels and flooding extent are used for model calibration/validation/verification.

Table 1.Some satellite mission and sensors used for hydrological studies

Mission	Sensor (s)	Application (hydrological)
Aqua	AIRS, AMSR-E, AMSU-A, CERES, HiRDLS, HSB, MODIS	Surface temperatures of land and ocean. (Flood mapping)
CryoSat	DORIS-NG, Laser Reflectors (ESA), SIRAL	Ice thickness (Applied also for near-shore mapping and inland water monitoring)
Envisat	AATSR, ASAR, ASAR (image mode), ASAR (wave mode), DORIS-NG, MERIS, MIPAS	Physical oceanography, ice and snow, (Ocean/ river water level altimetry)
ERS 1	AMI/SAR/Image, AMI/SAR/Wave,	Earth Resources, Physical oceanography (altimetry)
ERS 2	AMI/Scatterometer, ATSR AMI/SAR/Image,AMI/SAR/Wave, AMI/Scatterometer, ATSR/M	Earth resources, Physical oceanography (altimetry)
Jason 1	DORIS-NG, JMR, LRA, POSEIDON-2 (SSALT-2), TRSR	Physical oceanography (Ocean/River water level altimetry)
Jason 2	AMR, DORIS-NG, GPSP, JMR, LRA, POSEIDON-3	Physical oceanography (altimetry)

Jason 3	POSEIDON-3B Altimeter, DORIS-NG, AMR-2, GPSP, LRA, JRE	Physical oceanography (altimetry sea level rise, ocean circulation, and climate change
Radarsat 1/2	-Band SAR, X-Band SAR	Flood mapping/modelling
Sentinel 1	C-Band SAR	Flood mapping/ modelling
SRTM	C-Band SAR, X-Band SAR	Digital elevation models, flood modelling
SPOT 4	DORIS (SPOT), HRVIR, VEGETATION	Digital terrain models, environmental monitoring
SPOT 5	DORIS-NG (SPOT), HRG, HRS, VEGETATION	Digital terrain models, environmental monitoring
Terra	MODIS, MOPITT, MISR, ASTER, CERES	Physical processes, surface temperatures of land and ocean (surface water mapping)
Topex/Poseidon	DORIS, LRA, POSEIDON-1 (SSALT-1), TMR, TOPEX	Physical oceanography (altimetry)

Choice of suitable observed data can introduce subjectivity in the modelling process and subsequently increase uncertainty. Consequently, satellite data used to benchmark the model output accuracy can influence model calibration and validation (Stephens, Bates, Freer, & Mason, 2012). A review of types of satellite data used for flood modelling by Yan, et al., (2015) discusses satellite data accuracy and methods used for error reduction.

A review of applications of satellite remote sensing in surface water modelling, mapping and estimation is presented here and its limitations for surface water applications are also discussed. The review limits itself to water flowing within channels and coastal areas, and therefore excludes applications of satellite remote sensing for soil moisture measurement, rainfall estimation, rainfall/run off modelling and its associated routing estimations.

3.2 Overview of satellite data applications for surface water studies

3.2.1 SAR data applications

SAR data are useful for flood extent measurements even in cloud covered areas, and are therefore often used to make flood maps (Vermeulen, Barneveld, Huizinga, & Havinga, 2005; Horritt, 2006; Mason, Horritt, Dall'Amico, & Scott, 2007; Schumann, et al., 2007; Di Baldassarre, Schumann, & Bates, 2009; Long, Fatoyinbo, & Policelli, 2014). The variation of radar backscatter from different targets enables flood extent mapping. Several methods have been used to delineate the flooding extent from SAR data; e.g. utilization of multi-polarized Advanced SAR images, application of a statistical active control model, multi-temporal image enhancement and differencing, histogram thresholding/ clustering, radiometric thresholding, pixel-based segmentation, use of artificial neural networks, etc. (Long et al,. 2014). Multi-temporal image flood mapping involves acquiring flooding and non-flood images of the same area and combing them to get an image which indicates change by colours appearing in the image. A modification of the multi-temporal technique introduces an index that shows the changing areas (Skakun, 2010).

Sarhadi, et al., (2012) applied satellite stereoscopic images of Cartosat-1 to delineate flood hazard maps; the method used Rational Polynomial Coefficients to extract a high resolution DTM and detailed parameterization of the channel in Halilrud basin and Jiroft city in south-eastern Iran.

Segmentation threshold algorithms are used to delineate flood extents after a threshold has been manually chosen. Flood extent maps were created over four years of seasonal flooding in the Chobe floodplain, Namibia (Long et al,. 2014). 11 scenes of SAR data were enhanced using adaptive Gamma filtering (to remove speckles), and difference images created by subtracting from the reference non flood season image. The histograms of the difference images were then used to create thresholds separating flooded and non-flooded areas. The threshold for flooded areas was determined by subtracting the standard deviation multiplied by a coefficient Kf from the mean pixel value. For flooding under vegetated areas, the threshold was determined adding the standard deviation multiplied by a coefficient Kfv to the mean pixel value. The flood maps were then created using segmentation clustering in ENVI. Segmentation based on self-organizing Kohonen's maps (SOM) neural networks was used by Skakun (2010) to map flooding from five rivers in China, India, Hungary, Ukraine, Laos and Thailand. Training and testing of the neural networks were based on ground-truth data which enabled classification of water and dry land pixels.

Figure 3.1. NigeriaSatX image showing the confluence of the Rivers Niger and Benue at Lokoja, Nigeria. Satellite image water colours indicate differences in sediment load between the two rivers.

SOM produces a low dimensional representation of the input space that still preserves the topological properties of the input space. The method enabled automatic discovery of statistically salient features of pattern vectors, clustering and classification of new patterns. The resulting flood maps show an 85-95% classification rate compared with independent testing data; showing the applicability of the method for emergency flood mapping.

Interferometric phase difference between two SAR images is called the interferogram and includes signatures from topography, noise, displacement, atmospheric effects and baseline error. The advantage of phase changes in SAR interferometer data (INSAR) enables detection of change in the Earths land-use and land cover. This characteristic is very useful for identification of flooded areas over wetlands as used by (Dellepiane, de Laurentiis, & Giordano, 2004). The method, based on fuzzy connectivity concepts, automatically selected the coastline from two InSAR imagery using the coherence of the two images.

InSAR has been used to calculate the changes in water levels using satellite altimetry data for calibration (Kim, et al., 2009; Jung, et al., 2010). To obtain the displacement phase used to obtain the change in water height, all other signals are removed. The interferogram data gives the relative water level change between two locations. Where there is measured water level data (within acceptable radius) the relative water level change can be converted into the absolute water level change. Jung el al., (2010) used interferometric SAR data from JERS-1 to study change in water levels for the Amazon and Congo rivers. The data were acquired for the low flow and high flow seasons and processed using the 'two pass' method which includes flat earth phase removal and interferometric phase removal. Flooded vegetation, non-flooded areas and open water were differentiated based on backscatter 'noise floor' and 'mean interferometric coherence' of flooded and non-flooded areas. The temporal variation in water level dh/dt was obtained by converting the phase changes in imagery to water level referenced to the WGS84 datum using altimeter measurements from Topex/Poseidon. Using dh/dt to characterize the Amazon floodplain showed increasing dh/dt from upstream to downstream within a complex pattern of interconnected channels with distinct boundaries and varying dh/dt. The Congo River characterization of dh/dt showed a uniformity and limited connectivity between the river and the adjoining wetlands. Schumann et al., (2007) used Envisat ASAR data to identify spatial clusters of channel roughness in order to calibrate a HEC-RAS model of Alzette river flooding. ERS SAR data of the same event and an aerial photo of an earlier event were used for validation of the calibrated model and overall model performance was compared to measured high water marks at seven points during the flood event. The mean cross sectional water levels used for model evaluation were estimated from the intersection of ASAR flood extent boundaries with LIDAR DEM. At each cross section, ranges of channel roughness values are run in a Monte Carlo simulation and the CDF's of the values are generated; these CDF's are compared with a CDF of uniformly distributed model (where model functioning is same over the entire parameter space). The deviation of the individual CDF's from the CDF of uniform distribution give the measure of the parameter sensitivity, the sum of which show the local functioning of the model at that cross section. CDF's with similar error characteristics are grouped into clusters using k-mean clustering. The results showed that two clusters of roughness values are enough to measure the parameter sensitivity.

To utilise SAR data for flood depth estimation, methods have been developed that derive flood heights from flood extent data. The methods used combine SAR data with elevation data sources like DEMs, altimetry, and TINs. Mason, et al., (2007) and Schumann, et al., (2006), estimated the mean cross sectional water levels used for model evaluation from the intersection of SAR flood extent boundaries with LIDAR DEM. Schumann, et al., (2006) used linear regression and an elevation based model (REFIX) to convert SAR flood extent to heights and derived the flood water depth. Assuming a horizontal water height at cross sections, the water levels on the left and right banks were taken and used to subtract the floodplain DEM values to get the water height. The flood water line was then extracted using the regression equation: H= a.d+b, where a= slope of the regression line, d= downstream water distance, b= intercept. Using the cross sections as break lines and the flood water heights extracted, a TIN of the water heights at each cross section was produced. The flood water depth was derived by subtracting the DEM at the cross sectional interception points from the flood water height TIN. The result showed an RMSE of 18cm for a channel with no hydraulic structures and 31cm for a channel with many hydraulic structures and changes in slope. The study recommends that nonlinear regression/ piece wise regression can be used in the case of sudden changes in slope (due to hydraulic structures etc.) that cause the channel geometry to change.

Altimetry data from ENVISAT was combined with INSAR data from PALSAR and Radarsat-1 to compute absolute water level changes over the wetlands of Louisiana (Kim et al., 2009). Two pass INSAR method was used to check the two SAR images acquired at different times for phase differences. The ENVISAT altimetry data was used as the reference absolute water level change dh0/dt to compute the all the changes in water level over the domain. The results obtained for water level changes showed better comparison with the wetland gauge than with the channel gauge which had many levees interrupting the flow. Westahoff, et al., (2010) mapped probabilistic flood extents from SAR data by using the amount of backscatter and local incidence angles to create histograms that distinguish between wet and dry areas. The histograms were used to calculate the probability of flooding of every pixel.

Satellite data is used to calibrate hydrologic models especially in un-gauged catchments (Vermeulen, Barneveld, Huizinga, & Havinga, 2005; Sun, Ishidaira, & Bastola, 2009). Calibration of flood inundation models can be done using several model parameters, but the most sensitive parameter that shows a direct relation with water stage and therefore flooding extent and timing is the channel roughness (Schumann et al., 2007). Woldemicheal, et al., (2010) showed that for braided rivers where the hydraulic radius is obtained from indirect sources like satellite data, Manning's roughness coefficients can be used to minimize computed water level outliers. Roughness coefficient values to be used for calibration can be determined via flood modelling where the measured data are available.

Satellite based maps of flood extent have been used to calibrate flood inundation models either based on single or multiple flood events (Di Baldassarre, et al., 2009). Horritt (2006) calibrated and validated a model of uncertain flood inundation extent for the Severn River using observed flooded extent mapped from satellite imagery. Model accuracy was checked using reliability diagrams, and model precision was checked using an entropy-like measure which computes the level of uncertainty in the flood inundation map. The ensemble model outputs were compared with ERS and Radarsat data for calibration using the measure of fit. The results showed that the mapped flood extent produced only a modest reduction in the uncertainty of model predictions because the timing of satellite passes did not coincide with the flood event. Di Baldassarre, et al., (2009) showed that satellite flood imagery acquired during an event can be reliable for flood mapping. They used imagery of a single event covered by two satellite passes captured almost at the same time to develop a method to calibrate flood inundation models based on 'possible' inundation extents from the two imageries. Hydrodynamic flood model extents were compared with the satellite flood extent maps in order to calibrate the floodplain frictional parameters and determine the best satellite resolution for flood extent mapping. In spite of their different resolutions the result showed that both satellite imageries could be used for model calibration, but different frictional values have to be used in the model.

For un-gauged basins where hydrological data is inaccessible, satellite measurement of river width can be used for hydrological model calibration (Schumann, et al., 2013; Sun, Ishidaira, & Bastola, 2009). River width can be estimated from several sources of satellite data; making it more readily available than discharge or water level. Sun, et al., (2009) used measured river width from satellite SAR imagery to calibrate HYMOD hydrological model. The model calibration based on river width gave 88.24% Nash coefficient, with a larger error during low flow than high flow periods; implying its usefulness for flood discharge calculations. From the results, braided rivers showed lower errors for good Q-W relations from satellites. However, a small error in width measurement can lead to a large error in discharge estimation as the discharge variability was much larger than the width variability. Sun, et al., (2010) used the GLUE methodology to reduce this uncertainty in calibration of river width -to- discharge estimation with the HYMOD hydrological model. From 50000 samples of the parameter sets, 151 (Likelihood=RMSE values) succeeded as behavioural sets to be used in the model to simulate the measured satellite river widths. River discharge simulated with the successful parameters (Likelihood = Nash-Sutcliffe efficiency) gave good discharge simulation with a correlation $R^2 = 0.92$.

Model use in forecasting is affected by the propagation of the input uncertainties which make it less accurate. Data assimilation can be used to reduce the accumulation of errors in hydraulic models. Assimilation combines model predictions with observations and quantifies the errors between them in order to determine the optimal model and improve future forecasts (Mcmillan, et al., 2013). Types of assimilation techniques include Kalman filter (and its variations), particle filter and variational technique. Particle filter assimilation is a Bayesian learning system which accounts for input data uncertainty propagation by selecting suitable input data from randomly generated ones without assuming any particular distribution of their PDF (Noh, et al., 2011). Particle filter technique was used in studies like Matgen, et al., (2010), Giustarini, et al., (2011) where input data are in form of ensemble flow outputs of a hydrological model. In Giustarini, et al., (2011) to assimilate water levels derived from two SAR images of flooding in the Alzette River into a hydraulic model, 64 upstream flows were generated from an ensemble hydrologic model and used as the upstream boundary conditions. The most commonly used data assimilation technique however, is the Kalman filter which is a state-space filtering method which assumes a Gaussian distribution of errors. Vermeulen et al., (2005) used SAR derived flood maps and time series data to make flood forecasting more accurate through data assimilation. The assimilation process based on kalman filtering technique used adaptation factors to multiply the original model output and adaptation factor in order to generate a new parameter value. The process included calculation of water levels/discharge on the Rhine River by combining hydrologic modelling of the sub-basins and hydraulic modelling using downstream measured data. Data assimilation was done using measured water levels to determine the roughness coefficients which calibrate the calculated water levels. The model output water levels were compared with water levels derived from flood maps but because the natural flow of the channel or floodplain has been modified, good results were only obtained when the geo-referencing of the map is deliberately shifted or the flooding extent is exaggerated by adding some random noise over a large area of 7-12km. Barneveld, et al., (2008) applied the same method and models for flood forecasting on the Rhine River and produced good results of 10 day forecasts; therefore assimilating data for natural catchments results in better forecast model values. More information on hydrologic data assimilation techniques can be found in (Matgen, et al., 2010; Chen, Yang, Hong, Gourley, & Zhang, 2013; García-Pintado, et al., 2015).

3.2.2   Satellite altimetry data applications

Satellite altimetry (figure 3.2) works on the principle of return echo of pulses sent from the satellite nadir point and reflected from the surfaces of open water.

The height of the water surface is extracted from the distance between the satellite and the water body with reference to a local datum given as:

$$h = R - \left(c\frac{\Delta t}{2} \right) - \Sigma cor \qquad (1)$$

Where h= water level, R = distance between the satellite altimeter and the water body, c= speed of light, $\Delta t/2$ = two way travel time of radar pulse, Σcor = sum of corrections for ionospheric, wet and dry tropospheric, and tidal corrections.

This principle (figure 3.2) has its limitations as the accuracy of the data is affected by atmospheric conditions, sensor and satellite characteristics, and reflectance conditions (Belaud, Cassan, & Bader, 2010).

Although satellite altimetry was developed and optimized to measure ocean level changes (not rivers), it has been demonstrated as a source of data over large rivers and lakes (Tarpanelli, Barbetta, Brocca, & Moramarco, 2013; Jarihani A. , Callow, Johansen, & Gouweleeuw, 2013). Typical altimeter footprints are in kilometres; e.g. ENVISAT ranges from 1.6-10.8km, TOPEX/POSEIDON from 2.0-16.4km. Thus satellite altimetry data is used as the primary source of water level data in ungauged basins, and as a secondary data source to compare with measured data in sparsely gauged basins.

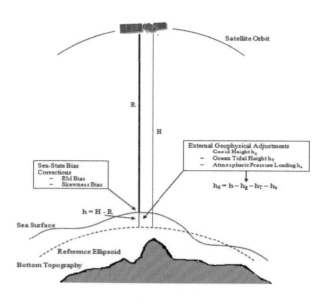

Figure 3.2. : An illustration of height measurement using satellite Altimetry.

Often times the selection of altimeter water level data to be used depends on the time and season of acquisition (Papa, et al., 2012). Data acquired during high flows give better measurements than low flow season data which usually have artefacts (in the form of islets, river banks, vegetation, etc.) that reduce the accuracy of the data in comparison with local gauge data. Analysing data over the Ganga-Brahmaputra Rivers, Papa, et al., (2012) got mean errors less than cm when high flow altimetry data were compared with measured data, but low flow data showed errors larger than 30cm. Siddique-E-Akbor, et al., (2011) used data from ENVISAT to compare with 1D HECRAS model output water levels in order to check for accuracy and ability to get the seasonal trend. The model was run for periods of available ENVISAT data and the output compared with the ENVISAT time series. The results showed RMSE ranging from 0.70-2.4m with the best correlation obtained during high flow seasons. The study suggest the use of calibrated hydrodynamic/ hydrologic model outputs to benchmark altimetry data in ungauged and poorly gauged catchments.

Virtual altimeter gauging stations are located at the intersection of satellite tracts with water bodies. Santos da Silva, et al., (2007) used virtual altimeter stations as water level data sources for ungauged catchments. They chose the median values of virtual stations that fell within river water bodies as water levels for the river and compared with measured values from gauging stations located within 20km of the virtual stations using weighted linear regression. In order to avoid comparing two areas with different hydrological conditions, a ratio χ was computed of the discrepancy between the ENVISAT master points and the linear regression and the uncertainties associated to the ENVISAT master points. The developed method enabled a comparison that produced regression coefficient greater than 0.95 between the ENVISAT and gauges series. Santos da Silver, et al., (2012) used 533 ENVISAT virtual stations and 106 gauging data to map extreme stage variations along 32 Amazon basin rivers and analysed for drought in the catchment. Using 2005 drought and 2009 flooding events as basis, data from 2002-2005 were analysed and time series of ENVISAT per virtual station were averaged to get monthly values. Values of the mean amplitude stage variation for the measured gauges showed good consistency with those of satellite altimetry and results for drought showed a range between -4 and 1m of anomalies. Getirana, et al., (2009) went even further and developed a rating curve of discharge values using virtual stations from ENVISAT for the upper part of the Branco River basin, Brazil. Virtual stations data were compared with nearby gauge data to check for seasonal similarity and trend, and those virtual stations with standard deviations <0.1m were chosen. The method used a distributed hydrological model to derive discharge values for the virtual stations. Model calibration and validation results showed good correlation with measured data, and the rating curve showed a 2.5-5% increase in bias when compared with rating curves from measured data. The calibration results were affected by rainfall data spatial distribution.

Although use of satellite altimetry for river stage monitoring is usually applied to large rivers with a few kilometres width (Papa, et al., 2012), altimetry data was used to estimate discharge in an ungauged part of the Po river basin (width:200-300m) using cross section data (Tarpanelli, Barbetta, Brocca, & Moramarco, 2013). They used a simplified routing model (RCM) based on upstream data, wave travel time and hydraulic conditions on two river sections to get the flow in the second river section. The results showed good agreement between simulated and insitu discharges, and gave lower RMSEs (relative to the mean observed discharge) than calculated results using an empirical equation also based on cross section geometry. Seyler, et al., (2009) used altimetry virtual stations to estimate river slope. The calculated river slopes were used to get the river bank full discharge, and the results compared well with gauge data. Lake water volumes were calculated for Lake Mead (USA) and Lake Tana (Ethiopia) using five altimetry data products: T/P (Topex/Poseidon), Jason-1, Jason-2, GFO (Geosat Fellow On), ICESat and ENVISAT (Duan & Bastiaanssen, 2013). The method used Landsat TM/ETM + imagery data to map the water surface areas using the Modified

Normalized Difference Water Index (MNDWI) method which enables robust extraction of water bodies from optical data (Zhang, et al., 2006). The calculated water surface areas agreed with in-situ measured data with an R^2 of 0.99for Lake Mead and 0.89 for Lake Tana with RMSEs of 2.19% for Lake Mead and 4.64%. The water volume was estimated using the lowest altimeter water level as the reference water level; this is then subtracted from all the other measurements to obtain the Water Level above Lowest Level (WLALL) to be used for volume estimation. Using regression analysis a relationship was established between the estimated water surface areas and the WLALL as A = f (WLALL) = aWLALL2 + bWLALL + c; where a, b, and c, are constants. The integral of this relation provides the Water Volume Above the Lowest water Level (WVALL). The estimated water volumes agreed well with in-situ water volumes for both Lake Mead and Lake Tana with R^2>0.95and RMSE ranging between 4.6 and 13.1%.

3.2.3   Optical satellite data

Depending on its contents, water reflects electromagnetic waves differently; pure clear water reflects differently from muddy water or water containing vegetation (floating or submerged). The amount of energy measured from the satellite sensor also depends on the bands used; blue band penetrates water up to 10m, red band is partially absorbed, and near infra-red band is totally absorbed. These sensor properties consequently affect the image, so that an image acquired using the blue band will measure reflectance from any submerged vegetation within its reach, while red/near infra-red images will show water as dark grey/ black respectively (Meijerink, Bannert, Batelaan, Lubczynski, & Pointet, 2007).

With the availability of more optical satellites with relatively low temporal resolutions globally, many scenes of archived data can be accessed and used for change detection studies and flood extent mapping in areas with little cloud cover. Penton & Overton, (2007) combined flood mask extents from LandSat ETM of four flood events with LIDAR DEM to produce water heights for the floodplain. The heights of the flood mask water points were used to interpolate a water height surface which was subtracted from the DEM to produce the inundation map. To check for water surface change, satellite microwave data from AMSR-E satellite was used to calibrate CREST hydrologic model using ratio brightness temperature measurements over water bodies and calibrated dry areas (Khan, et al., 2012). The AMSR-E detected water surface signal frequency was compared with gauge flow with a probability of exceedance <25% and showed good agreement. The output of model calibrated with AMSR-E detected water surface signal showed good agreement with observed flow frequency. Results of validation were equally good with high correlation between model results and observed flows with probability of exceedance <25%. The output of the model calibrated with AMSR-E detected water surface signal showed good agreement with observed flow frequency (Nash-Sutcliff coefficient of 0.90 and a correlation coefficient of 0.80).

Due to inaccessibility of the coastal terrain, many remote wetlands and swamps have few or no gauges, and are not covered by national gridding systems. As a result such areas are not included in topographic mapping projects; even where data is available the resolution is usually very coarse and not detailed (e.g. in Ezer & Liu, 2010). The morphology of coastal areas are affected by sediment supply, sea level change, littoral transport, storm surges, as well as hydrodynamics at the river mouths of deltaic areas (Kumar, Narayana, & Jayappa, 2010a). Tidal flat morphology for example, changes with the tidal cycle and this can affect navigation, coastal defence, fishing, etc. The monitoring and modelling of tidal flat morphology is thus important (Mason, Scott, & Dance, Remote sensing of intertidal morphological change in Morcambe Bay, U,K., between 1991 and 2007., 2010). Apart from natural causes, coastal areas are affected by human activities like sand mining, and construction of coastal infrastructure like ports, harbours, groins and other coastal defence systems.

Satellite data are used to study coastal morphological changes that affect the ecosystem and biodiversity of coastal areas. Kumar, et al., (2010a) studied the morphological changes in coastal parts Karnataka State, India using satellite and ancillary data. They calculated the rate of shoreline change over a ninety five year period (1910-2005) and used the results to predict future shoreline change rates to 2029. 25 LandSat TM imageries were used to map the tidal mudflats of Cooks Inlet Alaska by integrating with an inundation model (Ezer & Liu, 2010). The morphology of Cooks Inlet is such that, tidal floods move much faster than the ebbing period which moves very slowly; therefore areas at the far end of the mudflats take several hours before tidal waters lower. To study their morphology as a test bed for prediction of floods and its effects, mapping of these frequently flooded areas was done using the LandSat imagery to delineate water only areas, and show the range of shoreline data and water levels. The model results calculated the water depth and gave the estimated 3D topography of Cooks Inlet. Similarly, four LandSat TM imagery of the Ganges -Brahmaputra River mouth taken during low-flow and high-flow seasons were used by Islam, et al., (2002) to estimate suspended sediment concentration. The method used converted the digital numbers of the imageries to radiance values and subsequently to spectral reflectance and linearly related them to suspended sediment concentration (SSC). The SSC results showed higher distribution of suspended sediments during high discharge seasons when the turbidity zone moves further seaward reaching debts of 10m, than during low flow periods when the turbidity zone remains close to the shore. Yang & Ouchic, (2012) used 2000-2009 optical and SAR satellite imagery and insitu data of the Han estuary in Korea to study bar morphology by relating it with tides and precipitation using regression analysis. The results showed areas closer to the sea correlating bar size/shape with tides, and areas closer to the river mouths correlating with precipitation.

Optical satellite images of Sumatra Island were used to study post tsunami coastal recovery based on beach nourishment and sediment refilling. Liew, et al., (2010) used 1m Ikonos images of pre-tsunami, tsunami, and post tsunami periods to show that coasts affected by tsunamis naturally rebuild to their former morphological states in areas with little anthropogenic activity. The results showed straight beaches rebuilding few weeks after the tsunami, but recovery of barrier beaches and lagoons is much slower, enabling inland rivers and streams to directly discharge into the ocean. Thus, they concluded that due to the fast recovery of coastal features post tsunami, sedimentary deposits are better indicators of coastal geomorphology than tsunami events.

3.2.4   Satellite-derived DEM data applications

Satellite data provide topographic information in the form of digital elevation models (DEM's) generated from radar echoes of spot heights e.g. ASTER DEM, SRTM, and SPOT DEM. The most common and freely available DEM is the Shuttle Radar Topographic Mission (SRTM) DEM flown in February 2000 which covered 85% of the earth's surface. SRTM which was obtained through SAR interferometry of C-band signals is available in 30m and 90m spatial resolutions and an approximate vertical accuracy of 3.7m (Syvitski, Overeem, Brakenridge, & Hannon, 2012). The vertical accuracy of SRTM is higher in areas with gentle slopes than on steep slopes; on low-lying floodplains SRTM has shown less than 2m accuracy. More information on SRTM DEM accuracy can be found in (Yan, Di Baldassarre, Solomatine, & Schumann, A review of low-cost space-borne data for flood modelling:topography, flood extent and water level, 2015; Jarihani A. , Callow, Johansen, & Gouweleeuw, 2013). A comparison of SRTM data accuracy with 40m (topographic) contour data of Nigeria is shown in appendix A.

At the land-water boundary in areas with gentle slopes, satellite DEMs can be used to measure river stage when combined with high resolution imagery. Such combinations have been used in flood inundation mapping, although there is less accuracy in situations where the water edge is obscured by vegetation (Smith L. C., 1997). Syvitski, et al., (2012) adjusted SRTM data using ocean heights measured by the TOPEX/POSEIDON satellite altimeter to enable the mapping of floodplain zones. Advanced microwave Scanning Radiometer (AMSR-E) data provided brightness temperature measurements of the floodplain. The ratio of land area brightness temperatures to water area brightness temperature gave the discharge estimator; chosen dry areas were used as calibration areas for measurements over water covered areas. A rating curve of the ratio versus discharge was then used to extract the discharge values. Four floodplain zones were classified around the world from the 33 floodplains studied, namely: container valleys, floodplain depressions, nodal avulsions and delta plains. SRTM data measure surface level which over river channels is equivalent to water levels when the land water boundary is delineated. Jung, et al., (2010) used insitu (bathymetry and cross sectional) data and SRTM DEM water levels to derive water surface slope, and calculate the discharge of the Brahmaputra River. The cross sectional water level was obtained by fitting a first degree polynomial function to the SRTM data elevation. The average calculated discharge results when compared to insitu gauge reading gave a difference of 2.3%. Two DEM's of the Morecambe bay were used to determine the relative change in inter-tidal sediment volume above and below mean sea level(Mason, et al., 2010). The first set of DEMs was derived from satellite SAR imagery and the second set from LiDAR. By using the sea height as zero level the LiDAR DEM was normalized to the same height as the SAR DEM. The relative change in sediment volume was derived by subtracting the normalized LiDAR DEM heights from the SAR DEM. SRTM 30m data was combined with MODIS 500m water mask data to produce 30m static water masks of 2003 flooding along the Mississipi river (Li, Sun, Goldberg, & Stefanidis, 2013). The method involved using SRTM to mark the minimum water level from the MODIS water mask, which is then used to calculate the maximum water-level for that pixel using a water fraction relation. All SRTM 30m pixels with heights between minimum and maximum water levels are classified as water, and all those with heights higher than the maximum level are classified as dry. Consequently, the 500m MODIS water mask is integrated into a 30m water mask with the SRTM. The results gave detailed flood maps with the same flooding coverage as the MODIS water masks but enlarged 18 times. The flood maps were compared with Landsat TM images of the flood and showed over 94% match in water area coverage. Errors/ mismatch were found to be mostly around areas with trees and vegetation cover.

3.2.5   Gaps and limitations

As useful as satellite data applications have been in estimating surface water parameters, the measurements come with limitations due to sensor specifications/ errors, pre and post data processing techniques, calibration, measurement conditions, satellite distance from the targets, etc. Optical satellite data for example is limited to day time acquisition due to its dependence on sunlight, and is not very useful in areas perpetually covered by clouds because the target cannot be reached (Smith, 1997).

Since satellite data is used for calibration, its accuracy when compared with measured data is very important. Satellite data accuracy is estimated using different error measurement techniques (e.g. RMSE, Mean error), checking for correlation with measured data, or measuring the coefficient of determination (e.g. Tarpanelli, et al., 2013). There are multiple sources of error that can affect the data; for example the uncertainties in using satellite river width for calibration include: the satellite estimation of the river width, the power relation between the discharge and river width (which is an approximation of the conditions at a river cross section when there is no backwater effect) and the assumption of a stable/static river cross section. However these sources of uncertainty are lowest for the period of satellite data acquisition and increase with change in season and hydraulic conditions (Sun, Ishidaira, & Bastola, 2010).

3.2.5.1   SAR

The quality and usefulness of SAR data for hydrological studies depends on meteorological conditions (wind and rain), emergent vegetation, incidence angle and the polarisation mode used for data acquisition. Horizontal - Horizontal (HH) polarisation gives better results for flood extent mapping than Vertical - Horizontal (VH) and Vertical - Vertical (VV) polarisations. However, VH and VV polarisations are also useful since VV polarisation data highlight vertical features like vegetation, and VH polarisation data reflect the horizontal nature of the smoothed flood water (Schumann, et al., 2007). Another important factor for SAR data use in hydrology is the river size. Until the recent launch of CSK, RADARSAT-2, PALSAR, and TerraSAR-X, most available SAR satellites had large spatial resolutions which excluded smaller rivers from being captured; since it was difficult to delineate them in an image (Sun, et al., 2009).

Satellite SAR used for delineation of water extent has the limitation of floodplain vegetation being included and classified as water pixels; more so the height of the SAR waterline does not show the variation in water height with flow direction.

3.2.5.2   *Altimetry*

For river stage estimation and wetlands delineation, problems encountered with satellite altimetry data include: incorrect processing of radar echoes over rivers/lakes by satellite trackers, poor temporal resolution, and lack of information within the data about the atmospheric wet vapour content over lakes/rivers (Crétaux, et al., 2009). The errors recorded while using altimeter water level data can however be increased by incorrect choice of data; which frequently occurs when dry area data is retained within the data for computing water stages in low flow seasons (Santos da Silva, et al., 2007). The difference between altimeter and gauge measurements also increases with distance between the points, topography and river width (León, et al., 2006). When compared with gauge data, RMSEs of altimetry data measured over the Amazon have ranges from 30cm-70cm using data from ENVISAT, ERS2, and GeoSaT (Tarpanelli, et al., 2013),however at cross track situations where altimetry measurements are taken at the same location with a gauging station the difference can be <20cm (Seyler, et al., 2009). The accuracy of altimeter measurements over rivers is also affected by the river width and the morphology of the river banks so that data on narrow rivers and vegetated banks have lower accuracy (Papa, et al., 2012). Furthermore, the specifications of the altimetry system itself can affect quality of measurements; for example ENVISAT data have been shown to have lower RMSE compared to ERS2 data due to ENVISATs ability to switch frequency modes in response to change in terrain and its smaller bin width (Tarpanelli, et al., 2013).

3.2.5.3   *DEM*

The limitation of satellite DEM is in the data quality. DEM data needed for modelling and other analyses that require topographic data depends on the acquisition method, the data processing and the characteristics of the mapped terrain. Satellite derived DEMs have less vertical accuracy, higher bias and higher RMSE than other DEMS derived from airborne LIDAR and airborne IFSAR (Fraser & Ravanbakhsh, 2011).

In spite of their limited accuracy satellite DEMs have global or almost global coverage unlike airborne DEMs. Therefore they are useful sources of topographic data especially for low lying coastal areas with gentle slopes (Gorokhovich & Voustianiouk, 2006; Schumann, et al., 2008); and consequently applicable for inundation modelling (Karlsson & Arnberg, 2011). Figure 3.3 shows results of flood modelling undertaken for the Lower Niger River (Nigeria) using SRTM 30 and 90m. The Niger River overflowed its banks in 2007 and flooded a large part of the floodplain. MODIS satellite data was used to map the flood and provided the only reference record of the flood. Figure 3.3 shows that the model results are comparable with the MODIS flooding extent.

Generally satellite based DEMs are either generated from radar echoes of spot heights, or from SAR interferometry. However Mason et al., (2010) also derived DEMs from SAR images. The method involved using SAR water height to interpolate a set of waterlines, which were then used to produce a 50m DEM of the intertidal zone with an accuracy of 40cm. The method is however limited by the temporal de-correlation of the waterline heights.

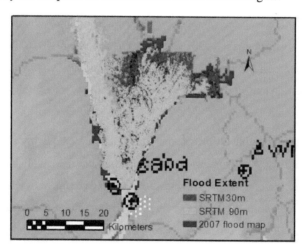

Figure 3.3.Model simulation result of flooding on the Niger (2007) using SRTM topographic data

3.2.6   Current data use strategies

Innovative methodologies are being introduced by scientists to better exploit satellite data to overcome the data limitations within present uncertainties. For example cloud filtering techniques have been developed that remove a high percentage of the clouds in optical data, thus adding to data availability. In terms of temporal limitations, combining MODIS data with its high temporal resolution with other types of satellite data is a technique that is now exploited more (Jarihani, et al., 2014). The technique generates new datasets that blend higher spatial resolution at the high temporal resolution of MODIS. When combined with DEM data for example, flood maps that provide daily information can be easily generated (Li, Sun, Goldberg, & Stefanidis, 2013). SRTM has been combined with MODIS data to generate a 250m water mask called MOD44W; because of the high temporal resolution of MODIS this product can be updated regularly to provide static water masks (Li, et al., 2013).

Use of Satellite SAR for flood extent mapping and model calibration can be improved through combination with other higher resolution data to increase precision in flood height determination. To improve the vertical accuracy of SAR waterline extent during floods, Mason, et al., (2007) used waterline data extracted from ERS-1 SAR corrected with 1m resolution LIDAR heights (along the Thames River bank) to calibrate a LISFLOOD model of flood extent. The output waterline when compared with waterline measured from aerial photos showed a lower root mean squared error than those obtained using SAR data only.

Satellite DEMs that are enhanced through vegetation smoothing or hydrological correction have shown lower errors compared with the original data (Jarihani A. , Callow, McVicar, Van Niel, & Larsen, 2015). Due to the availability of the hydrologically corrected SRTM DEM, a global static 30-m water mask has been generated which is very useful for flood detection especially in data scarce areas.

To improve the use of satellite altimetry data, interpolation methods have been developed to correct the data accuracy and precision by comparing the data with lakes and reservoir measurements. Thus the correlation with measured gauge data, range of RMSE and reduction in discrepancies have improved to levels >0.95 correlation during validation (Ričko, Birkett, Carton, & Crétauxc, 2012). Altimeter measurements over modified channels is however less reliable than that of natural catchments (Kim, et al., 2009).

The use of altimeter data is also limited by the poor temporal resolution of satellite altimeters; which range from days to several weeks. Belaud, et al., (2010) developed a method to interpolate river water levels in-between satellite observations in order to provide continuous data. The developed method used upstream ground station measurements and altimetry data as output to calibrate a propagation model by adjusting the satellite observed values. The propagation model used a transfer function to predict water level variations based on the relationship between the propagation times and water levels. The results were able to predict flood peaks during periods of no satellite coverage. Crétaux, et al., (2011) addressed the problem of data gaps by combining three sets of altimetry data (TOPEX/POSEIDON, ENVISAT1 and JASON2) with MODIS measurements of water extent to monitor wetlands and floodplains in arid/semi-arid regions. The MODIS data was used to classify the open water pixels whose relative values were then extracted from altimetry data. The results provided relative water heights, due to the low temporal resolution of the altimetry data sets. Altimeter data from ICESat was used to calibrate a large scale LISFLOOD-FP hydro-dynamic flood model of the Zambezi River, Mozambique (Schumann, et al., 2013). Eight in-channel water levels from ICESat from one altimeter pass were used for calibration of model output. The models with a mean bias within one standard deviation of the ICESat values were accepted as comparable with Landsat measured flooding extents. The results showed 86% agreement between the Landsat flood extent and the accepted model outputs; corresponding to mean distance of 1.42-1.60 km. After calibration the model upstream boundary was changed to forecast flow values in order to forecast downstream flooding. The results correlated with the baseline model, but showed that with a lead time of 5 days, better basin wide precipitation observations will enable flood forecasting on the Zambezi.

3.3 Use of high resolution insitu sampling

Data scarcity creates difficulties for hydraulic modelling in developing countries because river bathymetric data is unavailable in many cases. With only few gauging stations, the lower Niger River is one of the poorly gauged rivers of the world. Bathymetric data for most parts of the Niger River are not available but water depth data from dredging campaigns are available. The question arises if the data collected during such campaigns can be used to support flood modelling of the river. Dredging services typically start with a pre-dredging river survey during which hydrographic (e.g. water depth, bottom slope) and other survey data are acquired. Will a combination of such data with other data sources like terrain, flow and land-use, enable modelling of the river regime within acceptable accuracy? In this study such data is processed in combination with SRTM 30m DEM data to generate river cross-sections for the Lower Niger River; the cross sections are then used to model flooding. Use of high resolution data to modify SRTM DEM improves flood modelling in data scarce areas (Jena, Panigrahi, & Chatterjee, 2016).

3.3.1 Available data

The types of datasets available for the study include: hydrological, satellite imagery, satellite DEM, and hydrographic data (table 2). Measured hydrological data were available for Lokoja and Onitsha and were used as model boundary conditions. The slope data was measured at eleven points from Lokoja to Onitsha; the slope at the model downstream station at Aboh was calculated using measured distance between points and bed level data. Two scenes of Landsat 7 satellite imagery from December 2001 and January 2002 were also downloaded and used for study area characterisation and digitization of river bank extents. SRTM 30m resolution DEM obtained from the USGS EROS centre. The SRTM 30m over Africa was made available from September 2014.

Table 2.Types of datasets available

SI No	Type of Data		Data Source
1	Hydrological	Daily Discharge data	Nigerian Inland Waterways Authority (NIWA)
		Daily Water Level	
2	Hydrographic	Water depth	
		Slope	
3	Terrain	Digital Elevation data	USGS
4	Satellite Data	Image	

The 5-10m resolution survey water level and water depth data of the Niger River was obtained from Nigerian Inland Waterways Authority (NIWA) as described in sub-section 2.3.8 and shown in figure 2.6.

The SRTM DEM tiles were mosaicked and clipped to the extent of the flood plain; this includes the river flow area as digitized from Landsat imagery, and a buffer zone (figure 3.4). Cross sections on the DEM were checked for artificial sinks and sandbars by plotting the "Plot Profile" to determine consistency of DEM elevation values; and some anomalous values were observed (figure 3.4). These irregularities (verified by checking Landsat and Google Earth imagery for sandbar locations) were corrected in Arc GIS; this enabled manual water surface slope correction.

3.3.2   Utilizing dredging data for river cross section extraction and modelling

The surveyed water depth data were used to create a digital terrain model (DTM) for the Niger River. The data was converted from vector to raster using interpolation tools in ArcGIS. To find the best possible interpolation, three different methods were used: IDW- Inverse Distance Weighted, NN-Natural Neighbour, and Spline with barriers. The RMSE and MAE error of each interpolation were calculated by comparing measured and interpolated values; and results show that the IDW interpolated surface had less error and was thus chosen for the water depth raster creation with a cell size of 30x30m like that of SRTM DEM.

Bank lines of the Niger River were digitized from Landsat image and the position accuracy confirmed using, SRTM DEM and the outer edge of surveyed water depth transect lines. The river bank elevations were extracted from DEM using vector points added in the line shape file for both banks of the Niger River.

Since SRTM data values over water bodies represents the water levels, the channel bed levels can be obtained by using the simple concept that the river water level is the sum of the bed level and the water depth. Based on this, a DTM of the bed level was produced by subtracting the new water depths raster from the hydrologically corrected SRTM DEM. The DTM formed the 'bathymetry' of the Niger River in-channel and was used to obtain river cross section in-channel heights, while SRTM 30m provided the river bank elevation.

Transect lines were generated with 500 m interval within model reach (figure 3.4).The extent of transect lines in transverse direction was mainly based on the bank lines of both sides and created in such a way that transect lines were perpendicular with respect to river flow direction. Two Sobek 1D/2D hydrodynamic models were set up with the extracted cross sections (detailed model description in chapter 4).

The first model was run using cross sections extracted from SRTM DEM data; the model was calibrated using one month water level measurements at Onitsha (from December 2001 to January 2002). The water level data for Onitsha was obtained during the same dredging campaign that provided the water depth data. The second model was run using cross sections extracted from the created river bathymetry from SRTM-water depth.

Flood inundation models can be calibrated using several model parameters, but the most sensitive parameter that shows a direct relation with water stage and therefore flooding extent and timing is the channel roughness. Using Manning's n roughness values, the best model output with SRTM 90m cross sections was obtained with Manning's n values of 0.030. the result had a RMSE of 0.73 and a correlation coefficient value of 0.73. The second model with cross sections created from the modified SRTM DEM, was obtained with Manning's n value of 0.010; the result had RMSE of 0.025 and a correlation coefficient of 0.79 (figure 3.6).

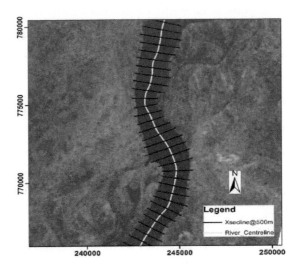

Figure 3.4. Cross sections on the Niger River; lines are perpendicular to the river centre line shown in yellow.

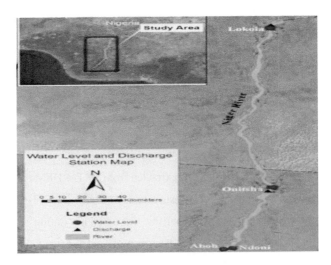

Figure 3.5. Dredging data coverage area on the Niger River. Onitsha water level data is used for model calibration

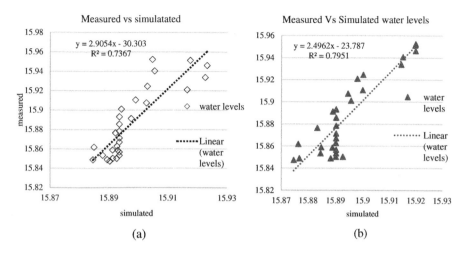

Figure 3.6. Model water level results comparison with measured data. (a) Model with SRTM DEM cross sections. (b) Model with modified SRTM DEM cross sections

Although the model results captured the flow regime and indicated where there is an increase in water levels as shown in the measured data, both are averaged across the cross section compared with the measured data; thus they both underestimate the highest water level and overestimate the lowest water levels at Onitsha. The model results with modified SRTM DEM however, simulates measured data with higher accuracy. Figure 3.7 shows the plots of the two model results compared to the measured data.

In 2012, Nigeria recorded the most extreme flood event in forty years when the Niger River recorded an over 500 year return period flood (Olomoda, 2012). This flood was captured by MODIS satellite and the flood extent mapped (NASA, 2012). With the availability of data from August to December 2012, the second model (with SRTM-DTM cross sections) was set up to simulate the flood extent. A 2D grid was added to the model setup to show the flooding extent, and the results successfully simulated the flooding extent as mapped.

In figure 3.8 (b) the model simulation result overlaid on MODIS satellite imagery, captures the flooding extent as well as estimates of the flood water depth.

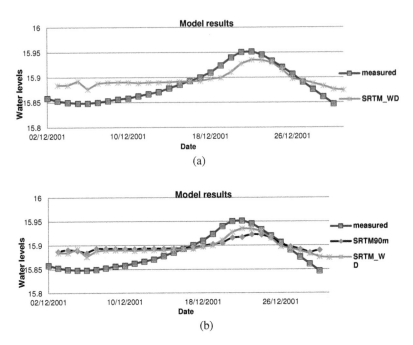

Figure 3.7. Flow model results at Onitsha (simulated water levels are averaged). (a) Model with cross sections from modified SRTM DEM. (b) Both model results are compared with measured data (b).

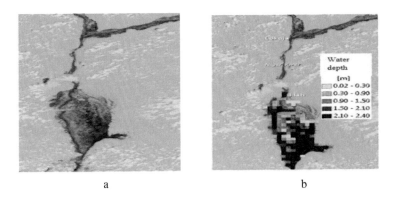

Figure 3.8. (a) Modis satellite image showing flooding on Niger River in 2012. (b) Model simulation results of 2012 flood overlaid on MODIS imagery.

The area simulated is a low lying depression (less than 40m a.m.s.l) that constitutes a large floodplain in the landscape. This result indicates that the modified SRTM DEM is useful and can improve flood modelling.

3.4 *Future direction*

Available literature show that efforts have been made to develop an empirical relationship between satellites derived surface water extents (including flooded areas) with river stage or discharge. Such a relationship has been established for braided rivers; for non-braided rivers the results have depended on the river system, thus inundation area can increase or decrease with stage. With better SAR missions such as TerraSAR-X- TanDEM-X formation, DEM data with good vertical accuracy are now available for better hydraulic flood modelling. TanDEM-X has 12.5m spatial resolution and produces less than 2m vertical accuracy (DLR, 2015). Although made for polar ice change estimation and monitoring, the high spatial coverage of Cryosat-2 is also being exploited for near-shore mapping and inland water monitoring (Villladsen, Andersen, & Stenseng, 2014). Cryosat-2 which operates in SAR and interferometric modes, has a drifting orbit and therefore (unlike all the other satellites) has little repetitive data (since repeat cycle is 369 days). Its high spatial density coverage makes it good for hydraulic modelling (and all its evaluations have produced good results). With successful use of Cryosat-2 data to obtain river water levels and topography, the use of drifting orbits is being proposed as more suitable for river water surface topography mapping, derivation of river profiles and building of pseudo time series (Bercher1, Calmant, Picot, Seyler, & Fleury, 2014).

Other satellite products that improve the accuracy of satellite data based research in hydrology include: Cosmo-SkyMed from the Italian Space Agency, RadarSat2 from the Canadian Space Agency, and Sentinel-1 from ESA (Schumann, Bates, Neal, & Andreadis, 2015). Others are Global Change Observation mission-water (GCOM-W) from Japan Space Agency (JAXA), Global Precipitation Measurement (GPM) from JAXA /USA, Soil Moisture Active Passive (SMAP) from USA.

To improve quality of satellite SAR and topographic data, new satellite missions with higher precision instruments are being planned. One of such missions is the Sentinel constellation that will consist of seven satellites; two of which (Sentinel 3 and 6) are especially dedicated to hydrological purposes. Sentinel 1 is already in orbit and undergoing calibration; it has a C-band SAR instrument to continue present C-band data provision. Sentinel 3 is planned to provide fast data for flood emergencies, therefore it has three instruments one of which is a dual-frequency (Ku and C band) advanced Synthetic Aperture Radar Altimeter (SRAL) that will provide accurate topographic data of oceans, ice sheets, sea ice, rivers and lakes (ESA, Sentinel

3, 2015). Sentinel 6, which will complement the Sentinel 3 data, will carry on board a high precision radar altimeter. RADARSAT constellation, a new Low Earth Orbit (LEO) C-band SAR mission is under development by the Canadian space Agency (CSA). The constellation which will have several operating modes will provide interferometric SAR data that can be used for wetlands and coastal change mapping, flood disaster warning and response with resolutions 3, 5, 16, 30, 50 and 100m (Canadian Space Agency (CSA), 2015).

Other upcoming satellite missions like Surface Water & Ocean Topography (SWOT) made especially to survey global surface water have specifications that will enable better use of satellite data in hydrology. SWOT which uses a wide-swath altimetry technology will also observe the fine details of the ocean's surface topography, and measure how water bodies change over time with repeated high-resolution elevation measurements. The mission, scheduled to be launched in 2020 is an international collaboration between the US National Aeronautics and Space Agency (NASA) and Centre National E'tudes Spatiales (CNES) of France; supported by the Canadian Space Agency (CSA) and the UK Space Agency (UKSA) (Pavelsky, et al., 2015). Other products of international cooperation that will support hydrological research are the JasonCS/Sentinel6, and the CFOSAT. CFOSAT is a French-Chinese joint satellite mission scheduled for launch in 2018. The satellite which will carry to main instruments, will monitor ocean surface wind and wave as well as other ocean and atmospheric science and applications. The JasonCS/ Sentinel6 mission will continue the work from Jason 3 by measuring sea surface height, wave, wind speed, and will provide useful data to monitor sea level rise, coastal areas modelling of oil spills, forecasting of hurricanes etc. Like Jason 3 it carries instruments to precisely detect sea level change; it combines GPS, radar altimetry, and a microwave radiometer to produce data within 1cm accuracy every 10 days (NOAA, 2015). Jason-CS / Sentinel-6 satellite bus will be based on CryoSat, and will be capable of mitigating space debris. The mission is scheduled to be launched in 2020, and the satellite will carry a new altimeter, the Poseidon-4; which will be able to provide simultaneous pulse-limited waveforms and Full Rate RAW waveforms that allow SAR processing on-ground (CNES, 2016). Jason-CS / Sentinel6 is jointly owned by US National Oceanic and Atmospheric Administration (NOAA), CNES-France, European Organisation for the Exploitation of Meteorological Satellites (EUMETSAT), European space agency (ESA), EU, and US NASA/JPL.

3.5 *Conclusion*

Hydrological data collection requires deployment of physical infrastructure like rain gauges, water level gauges, as well as use of expensive equipment like echo sounders. Appropriate model development requires sufficiently accurate representation of the river geometry as it critically influences the quality of the modelling results. Unavailability of high resolution topography and river cross-sections data are the prime concerns for developing hydrodynamic models (Jena, Panigrahi, & Chatterjee, 2016). To study ungauged river basins in data scarce areas therefore, several methods have been used by researchers; including use of alternative data sources.

Satellite remote sensing provides a source of hydrological data that is unhindered by geopolitical boundaries, has access to remote/unreachable areas, and provides frequent and reliable data (Jung, et al., 2010) This review shows the accuracy of satellite data use in hydrology (for river/ coast mapping, modelling, calibration, and parameter estimation) continues to improve. Use of satellite data has continued to advance due to easier data availability, improvement in sensor capabilities and improvement in knowledge of and utilization of satellite data, as well as expansion of research topics.

Dredging activities provide another source of hydrological data which can be used for flood modelling. Although this data source does not provide continues data as it is limited by project duration, the data obtained can be useful for simulating river conditions and for calibration and validation of flood models in data scarce areas. In this thesis, water depth data was used to create a DTM that was used to adjust former bathymetry created from SRTM 90m water levels. The new bathymetry created lowered the cross section dimensions and the output of flood models run with the new cross sections better simulated the river flow regime; results had lower errors compared with models run with SRTM only cross sections.

4

Modelling complex deltas

in data scarce areas[2]

[2] This chapter is an edited version of the following publications:

Musa, Z. N., Popescu, I. & Mynett, A., 2016. Approach on modelling complex deltas in data scarce areas: a case study of the lower Niger delta. Procedia Engineering, 104, 656 - 664. http://dx.doi.org/10.1016/j.proeng.2016.07.566

Book chapter: Musa, Z. N., Popescu, I. & Mynett, A., 2015. Uncertainty Analysis of Hydrodynamic Modelling of Flooding in the Lower Niger River under Sea Level Rise conditions In book: Advances in Hydro-informatics, Edition: 2364-6934, Chapter: 13, Publisher: Springer Singapore, Editors: Philippe Gourbesville, Jean A. Cunge, Guy Caignaert, pp.189-202

Natural processes consist of a combination of complex sub-processes and conditions. Models aim to represent the natural phenomena through simplified approximations of the physical processes in a reliable way. Flooding has many consequences including: destruction of properties/farmland, loss of human/animal life, inundation of dry land, contamination of surface water, etc. These effects of flooding can lead to loss of livelihood, poverty, economic depression, change in biodiversity, and water-borne disease (Bariweni, Tawari, & Abowei , 2012).

Flood modelling is formulated in dimensions representing different types of flow simplified for practical purposes. Hydrodynamic modelling of flooding in river channels and floodplains is based on the principles of continuity and momentum. For a control volume in an open channel under unsteady flow, the principle of continuity states that the mass flow in minus the mass flowing out should be the same as the change of volume in time. The momentum equations represent the balancing of forces acting on a water control volume. Depending on the flow channel properties, averaging is done such that flow can be modelled in one dimension (1D, e.g. flow in a pipe, stream), two dimensions (2D, e.g. flow in a shallow lake, coastal waters), and three dimensions (3D, e.g. wind driven currents on open water). To model river dynamics and the effects of floods, it is important to simulate the flow process and to represent the routes used by the river, rainfall and runoff. Consequently measurements of flow discharge/water levels, rainfall, run off and topographic information are important for flood modelling. However, data availability is a challenge for many researchers especially in developing countries where many catchments remain ungauged or sparsely gauged. The Niger delta is one of the low-lying deltas that will be affected by rising sea levels, but it lacks essential data for hydrologic and hydrodynamic modelling. In this chapter two types of modelling approaches are used to study the effects of SLR on the Niger delta through flooding and inundation. The models are run to simulate possible effects of SLR on the Niger delta by 2030, and 2050, and 2100.

The chapter is thus divided into two parts: the first section covers the possible effects of SLR on river dominated parts of the Niger delta. With river floods occurring more frequently than before, the aim of this section is to study the interaction between the strong hydraulic currents flowing from the upstream fresh water zone and the flooding in the coastal zone with future sea level rise.

Section 2 simulates the effects of coastal flooding on the wave dominated parts of the Niger delta under SLR conditions. For most coastal areas the possibility of coastal flooding in a year are quite high. High sea levels can cause inundation of coastal areas; e.g. parts of the Mississippi delta and Black River marshes in the US have already been submerged by rising sea levels (Titus, et al., 2009). Inundation due to sea level rise (SLR) will have a great effect if areas under fresh water regimes are turned into salt water swamps or wetlands. River deltas are the most susceptible coastal areas to inundation from SLR due to their natural tendency to subside in response to reduced sediment supply from upstream.

4.1 Effects of river flooding on coastal areas under sea level rise conditions

SOBEK hydrodynamic modelling tool, is used to model river flooding under SLR conditions. SOBEK is based on the full 1D St Venant equations for unsteady uniform flow in one dimension; thus the model assumes that all variations of velocity in the vertical directions and across channel are negligible, giving rise to uniform flow in all directions.

4.1.1 Methodology

Measured flow discharge values for 1998, 2005, 2006 and 2007 for Lokoja gauging station are used for the upstream boundary (figure 4.1). Based on the river flooding records, data for 1998, 2006 and 2007 represent flood year data, while 2005 data is normal flow data. The flood events of 2006 and 2007 resulted from high magnitude rainfall in the basin between July and September (Brakenridge, Kettner, Slayback, & Policelli, 2007), while the 1998 flood was due to heavy rainfall modified by a dam break from upstream which caused flash flooding in the channel in October (NDRMP, 2004a). Based on these data availability for different flow conditions in the Niger River, five modelling scenarios were simulated to evaluate the interaction of Niger River flooding with downstream rise in sea levels through effects on flooding extent, flooding time and water depth. The modelling scenarios were:

Scenario 1: Sea level rise with normal year flow from upstream,

Scenario 2: Sea level rise with a flooding year flow from upstream,

Scenario 3: Sea level rise with flash floods from upstream,

Scenario 4: Sea level rise with subsidence and flooding year flow from upstream,

Scenario 5: Sea level rise with subsidence and flash floods from upstream.

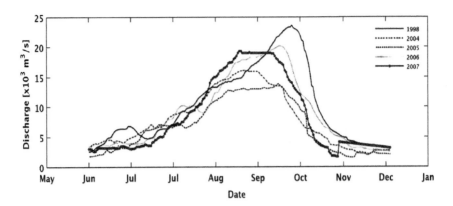

Figure 4.1. Hydrographs of flow used for model upstream boundary at Lokoja.

Measured subsidence values in the Niger delta show that the Niger delta is subsiding due to oil and gas extraction. Measured subsidence levels at active oil extraction sites range from 25 - 125 mm/year, while at settlement areas the range is 7-9 mm/year (Abam, 2001; Agabi, 2013).

For all models, tidal data were used for downstream water level boundary conditions[3]. SLR values used were 0.019m and 0.035m for the years 2030 and 2050 respectively obtained from (IPCC, 2013; Brown, Kebede, & Nicolls, 2011). Thus, with subsidence =7mm, SLR for 2030= 0.14m (7mm x period 2013-2030= 17years), SLR for 2050 = 0.29m (7mm x period 2013-2050 = 37years); with subsidence = 25mm, SLR 2030= 0.44m, and 2050=0.96m. Model simulations with subsidence were set with a Tolerance level of 0 – 0.2.

2D grids are generated for the floodplain from the Shuttle Radar Topography Mission (SRTM) DEM. All models in this thesis are referenced to mean sea level.

---

[3] These are Mean Higher High Water (MHHW) levels obtained from mareograph generated tide tables available from http://www.tides4fishing.com/tides. The mareograph prediction is based on 21 tide gauges in the Niger delta; the predicted data have a correlation of 99% with UK tide tables. There was no other source of tidal data (e.g. from the Nigerian Government) for the Niger delta coast.

4.1.1.1 Model Setup

A 1D and a 1D/2D hydrodynamic SOBEK models of flooding on the Niger River were set up (figure 4.2) with discharge data as upstream boundary conditions and tidal water level data as downstream boundary conditions. The sobek1D flow model was set up to run from June to November of each year with an initial water level of 0.1m; average seasonal low flow value obtained from river Niger records. The models are calibrated using Manning's roughness coefficient (from 0.03-0.05) and verified based on flood map of 2007 from Dartmouth flood observatory (Brakenridge, Kettner, Slayback, & Policelli, 2007). After calibration, the Manning's roughness coefficient with best results was 0.035 for the main flow channel and 0.045 for the channel sides. Since the Niger river bifurcates into the Forcados and Nun rivers at downstream end, the network (figure 4.2 has 48 flow cross sections, 1 upstream boundary located at Lokoja with discharge values as boundary condition, and 2 downstream boundaries located at the mouth ends of river Nun and Forcados with tidal data as boundary conditions. Calculation points[4] were set for every 150m along the reach.

The 1D/2D overland flow model is set up with the 1D flow model coupled with a 2D grid, to enable exchange of data between the 1D channel and the 2D grid. The 2D grid represents the floodplain and overlays the 1D channel. During simulations water flows down the 1D channel and only enters the 2D grid when there is an overflow of the river bank; this triggers the 2D overland flow model to start running.

4.1.2 Results and Discussion

To simulate the scenarios the boundary conditions are varied downstream at the mouths of rivers Forcados and Nun using sea level rise (SLR) values. First, the models were setup to replicate the past flood events, and then modified to simulate flooding under SLR conditions. Probability based uncertainty analysis is used to determine the uncertainty range of the modelling output which show the possible effects of SLR on the Niger delta. Pure random sampling is used to generate values from the projected 0.019 m - 0.35 m SLR (by 2030- 2050).

---

[4] Calculations are done at each calculation point in the Sobek network. Calculation points are the numerical grids used for simulation. At these grids the continuity and momentum equations are solved using the staggered grid convention by which the water levels are defined at the calculation points and nodes and the discharges are calculated at the reaches (Deltares, Sobek 1D/2D modelling suite for integral water solutions, 2013).

Since the Niger delta is subsiding due to oil and gas extraction, computed subsidence levels (0.14 m – 0.44 m by 2030, and 0.29 m - 0.96 m by 2050) were added to SLR values and modelled under scenarios 4 and 5. 25 1D simulations of each scenario are run using the random values generated. The simulations have a period of June to November of each year, and hourly time steps (giving an average of 3727 time steps). We use uniform distributions for the SLR and land subsidence values over the entire study area.

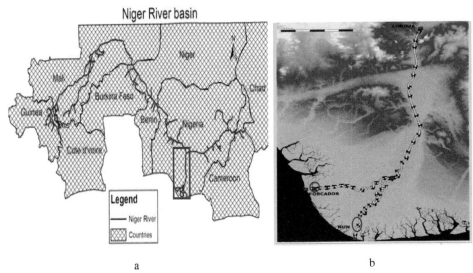

a   b

Figure 4.2. (a) The Niger River basin showing the tributaries and the member countries; the model reach is shown within the box. (b) Flow network from Lokoja set up with 48 flow cross sections and three flow boundaries. Locations of cross sections used for analysis are marked with red cycles.

1D River flow modelling

The 1D simulation results showed the possible effects of SLR on the flood events in the Niger River through extension in flooding time and increase in flooding extent. Detailed descriptions of the model results are given in the following subsections. An example of 1D water depth result at nodes is shown in appendix B.

Scenario1: SLR with normal flow (2005 hydrograph)

Figure 4.3 shows the result of twenty five simulations of the 1D model under normal flow conditions with SLR (ranging from 0.14 m - 0.35m) for the Nun River. Results indicate that with higher SLR values, initial water depths in the channel (in June) are much higher than the values without SLR. This results indicates that rise in sea levels will cause the flooding of areas around the Nun River which under 'normal' conditions does not overflow its banks in the rainy season.

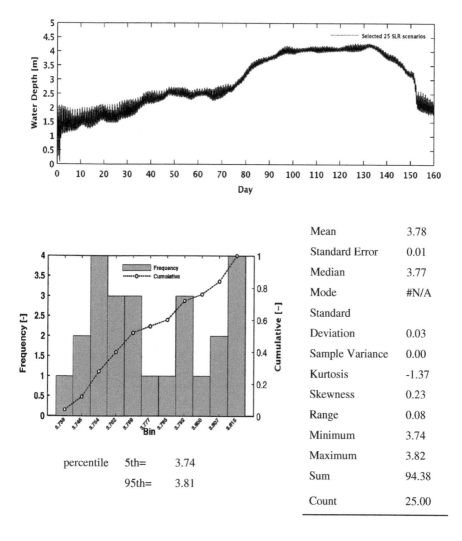

Figure 4.3. Nun River: 2005 'normal flow' with downstream SLR (above). Histogram and analysis of the simulation result (below).

Furthermore, analysis results shown in figure 4.3 indicate that with SLR and normal flow, the water depth will range between 3.74-3.82±0.01 with a 95% probability that it will not be higher than 3.8m above mean sea level. This implies that with 'normal' flow conditions and SLR (up to 0.035m), there will be some flooding due to the change in sea level, but water depths at the downstream end of river Nun will not significantly change between 2030 and 2050.

Scenario2: SLR with flooding year flow from upstream

The simulation results for 2006 flood hydrographs in figure 4.4 shows harmonized computation of water levels with little difference between the 25 computed values. The results analysis and histogram (figure 4.4) implies that at the downstream of Forcados River all levels of SLR will produce high water levels between 7.44m-7.47m.

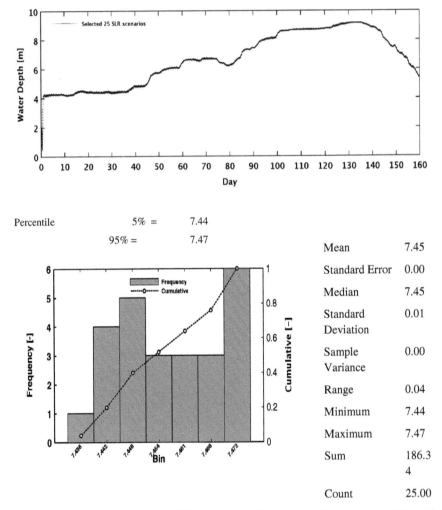

| Percentile | 5% = | 7.44 |
| | 95% = | 7.47 |

Mean	7.45
Standard Error	0.00
Median	7.45
Standard Deviation	0.01
Sample Variance	0.00
Range	0.04
Minimum	7.44
Maximum	7.47
Sum	186.34
Count	25.00

Figure 4.4. 2006 flood with downstream SLR for Forcados River (above). Histogram and analysis of the simulation result (below).

The standard error for the computation is 0.0 and there is only a 5% probability of the water levels being higher than the maximum water level.

Scenario3: SLR with flash floods from upstream

The simulation for SLR with flash flooding was done using the 1998 flood data. The results for SLR values of 0.019m by 2030 show no significant effects

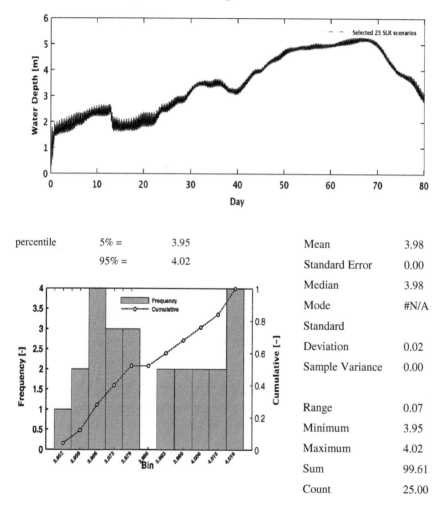

percentile	5% =	3.95	Mean	3.98
95% =	4.02	Standard Error	0.00
		Median	3.98
		Mode	#N/A
		Standard	
		Deviation	0.02
		Sample Variance	0.00
		Range	0.07
		Minimum	3.95
		Maximum	4.02
		Sum	99.61
		Count	25.00

Figure 4.5. 1998 flood on the Nun River with SLR. Histogram and analysis of the simulation result (below).

For SLR of 0.035m by 2050 however, the results show continued flooding in the channel from July such that the recession in water level in August (as in the case for low SLR) still allows enough water to flood the area. Thus the hydrographs in figure 4.5 show that the water depth during the August recession period (boxed area) is high at higher SLR values. The histogram for the 1998 flooding at river nun shown in figure 4.5 show very little difference between the minimum and maximum water levels, with 90% of the water levels lying between the minimum and maximum values. Thus with high magnitude SLR and flash flood, increase in water levels in the Nun River will last for a significant length of time.

Scenario 4: SLR with subsidence and flooding from upstream

To simulate SLR with subsidence, 25 random numbers were generated between 0.14 – 0.96m and added to downstream tidal water levels with SLR projections. The models were set with a Tolerance allowance of 0 – 0.2. 2007 flood data was used as upstream boundary condition and the simulation results with SLR and land subsidence shows high water depths at the mouth of River Forcados.

The 25 hydrographs generated show distinct water depths, which maintain an almost constant value for all time steps at higher subsidence levels. In figure 4.6 there is no drop in water depths in the entire simulation period for water depths higher than 8m. From the analysis hydrograph in figure 4.6, there is 90% probability that the water levels at downstream of Forcados will be between 7.8m -14.3m with land subsidence and SLR. The mean water level is 10.46m; which implies that with subsidence the probability of inundation in the Niger delta is very high. The standard error for this simulation is high with a standard deviation of 2.37m.

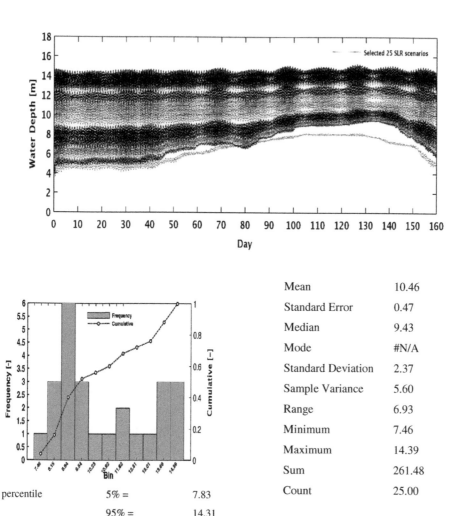

Mean	10.46
Standard Error	0.47
Median	9.43
Mode	#N/A
Standard Deviation	2.37
Sample Variance	5.60
Range	6.93
Minimum	7.46
Maximum	14.39
Sum	261.48
Count	25.00

| percentile | 5% = | 7.83 |
| | 95% = | 14.31 |

Figure 4.6. 2007 flood on the Forcados River with SLR and subsidence (above). Histogram and analysis of the simulation result (below).

Scenario 5: SLR with subsidence and flash flooding from upstream

The 1998 flash flood with land subsidence shows increase in water levels indicating the flash flood even at higher levels (figure 4.7). Results of the 1998 flood with SLR and land subsidence shows a doubling of water depths in flooded areas around the Forcados River. This is indicated in the analysis (figure 4.7) as the recession in water level (boxed area) is absent at water levels higher than 8m. The histogram shows a single spike at 7.07m indicating maintenance of high water levels for 92% of the simulations with 5% chance of water levels surpassing 8.93m. The error for this simulation is high with a standard deviation of 0.84m.

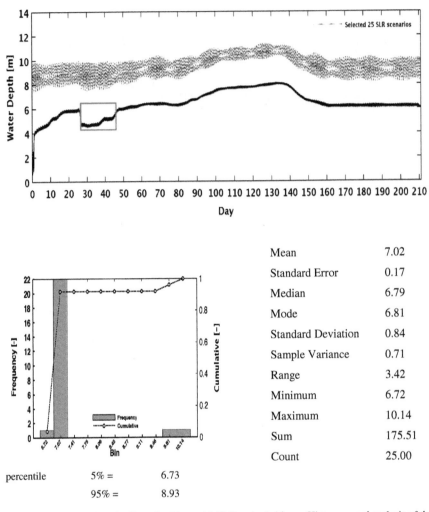

Mean		7.02
Standard Error		0.17
Median		6.79
Mode		6.81
Standard Deviation		0.84
Sample Variance		0.71
Range		3.42
Minimum		6.72
Maximum		10.14
Sum		175.51
Count		25.00

percentile	5% =	6.73
	95% =	8.93

Figure 4.7. 1998 flood on the Forcados River with SLR and subsidence. Histogram and analysis of the simulation results (below).

1D model results comparison

From the different scenario results, there are differences in flood extent and change in flood arrival time between models with SLR and those without. Table 3 compares the 1D results obtained for 2005, 2006 and 1998 for the different scenarios.

1D/2D Overland flow modelling

The 1D/2D overland flow model is a coupled model, therefore it is based on same settings as the 1D flow model coupled with 2D grids located at flooding locations identified from the 1D model run.

Table 3. Analysis of 1D simulation results

YEAR/ TYPE OF FLOW	DIFFERENCES IN SIMULATION RESULTS	
	NO SLR vs SLR	NO SLR vs SLR +SUBSIDENCE
2005/ *Normal year flow*	Forcados river: with SLR, flooding occurs one month earlier (July 15th instead of 20th August). Nun: flooding of coastal areas comes one month earlier (June instead of July) and upstream areas are flooded.	With SLR + subsidence, floods occur much earlier (first week of July in Forcados) and mid-June in Nun river.
2006/ *Flood year flow*	With SLR, flooding extends further upstream of the Forcados river past the bifurcation (up to 220km upstream in 2030 and 300km upstream in 2050). Floods also arrive one week earlier in 2050. Nun river: flood extends up to 250km upstream by 2030 and along the entire channel by 2050.	SLR+ subsidence by 2030 shows increase in flooded areas up to 350 km upstream for Forcados river and Nun river shows flooding along the entire channel. Floods also arrive 10 days earlier in the Nun river.
1998/ *Flood year flow + flash flood*	SLR by 2050 showed continued flooding in the channel from July 20 with no recession in water level in early August as was the case for no SLR.	With subsidence there is continues flooding in the channel from July to October.

Roughness coefficient obtained by calibration gave 0.045 for the 2D grids and 0.035 for the river channel. The 1D/2D overland flow simulation (e.g. figure 4.8) was used to check for changes in water depth, lateral flood extent and location of change for the most downstream coastal areas (closest to mouths of rivers Forcados and Nun). An overlay of the scenario results exported into ArcGIS show changes in flooding extent and water depth as discussed below. Another example of 1D/2D flooding result on Forcados and Nun rivers is shown in appendix B.

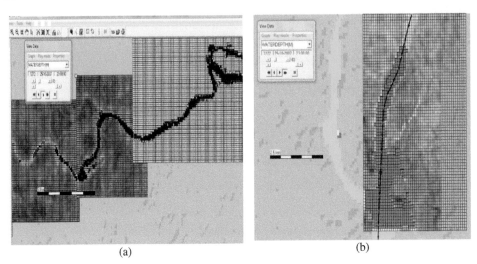

(a)   (b)

Figure 4.8.SOBEK 1D/2D result of 2007 flooding along the Forcados River (a), and downstream Nun River (b). Blue pixels show water, and each grid is 100 x 100 m2.

Nun River

The result for the most downstream areas around Nun River showed no differences in flooding extent or in water depth between all models with and without SLR by 2030 and 2050; i.e. with normal flow year (2005), flooding year flow (2007), and 1998 flooding year flow with flash flood. This result implies that at the downstream end of Nun River, the effects of SLR (up to 0.96 m) will not be significant on the river water depth and the flooding extent.

Forcados River

For downstream end of Forcados River, the results showed no difference in lateral flooding extent between the normal flow year/2007/1998 floods, and all ranges of SLR for 2030 and 2050 (0.14 m – 0.96 m). However water depths showed increases up to 1m in parts of the river even for models with 2005 normal flow year (figure 4.9 a). For models with 1998 (flash flood data) there is significant change in water depths with SLR; all flooded pixels show a doubling of water depth in the downstream river grid from 0 to1m and 1 to 2m for the years 2030 and 2050 (figure 4.9 b).

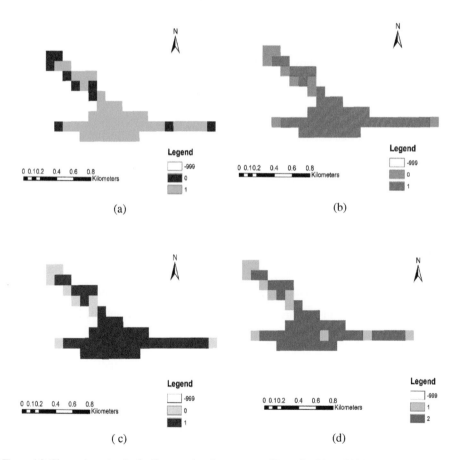

Figure 4.9. Change in water depths (in metres) at downstream of Forcados River. 2005 normal flow (a), 2005 normal flow with SLR (b), 1998 flood year flow with flash flood (c), 1998 flash flood with SLR(d).

This results imply that the effects of SLR downstream of the Forcados River will be noticeable and significant for river water depth; especially when there is a Dam break from upstream.

4.2 Effects of coastal flooding

For most coastal areas the possibility of coastal flooding in a year are quite high. Coastal flooding is caused by storms coinciding with high tides, by ocean surge, by high river flooding due to precipitation upstream (and locally), and sometimes by under ocean earth quakes known as tsunamis. Coastal flooding resulting from high river flood might occur as a result of flash floods, or when high sea levels impede the draining of river waters into the ocean. High sea levels can cause inundation of coastal areas; e.g. parts of the Mississippi delta and Black River marshes in the US have already been submerged by rising sea levels (Titus, et al., 2009). Inundation due to sea level rise (SLR) will have a great effect if areas under fresh water regimes are turned into salt water swamps or wetlands. River deltas are the most susceptible coastal areas to inundation from SLR due to their natural tendency to subside in response to reduced sediment supply from upstream.

Vulnerable coastlines make it easy for sea water to penetrate into inland areas, so that the effects of coastal floods, storm surges are felt several kilometres from the coast. The coastline of the Niger delta stretches for 450 km with different physical properties like geomorphology, topography, slope, longshore current, erosion rate, and tidal height. The coastline (with only a single location showing existence of groins), is only protected by its natural mangrove forests against storm surges. More so, out of the 320 settlements along the 450km coastline very few have canals (mostly located settlements with oil and gas facilities), dykes and water channels; and where these exist, they are built through local efforts with no clear compliance with design standards (NEST, Reports of Research Projects on Impacts and Adaptation, 2011).

A study of the vulnerability of Niger delta coastline, found the eastern end to be one of the most vulnerable to SLR (Musa, Popescu, & Mynett, 2014a). For this study the Bonny and new Calabar rivers located on the eastern end of the Niger delta are modelled. The two rivers are separated by a low lying floodplain but are hydrologically connected via various channels forming a single system. The area has very low topography, virtually absent fresh water input, two major estuaries (containing the Bonny-Calabar river system and a smaller estuary formed by a distributary of the Bonny River) and the highest tidal range in the Niger delta (Briggs,

Okowa, & Ndimele, 2013). The effects of SLR on coastal flooding for this area (figure 4.10) is modelled here.

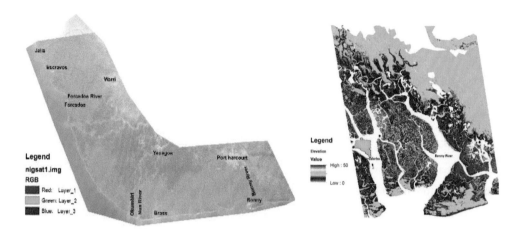

Figure 4.10. Modelled area at the eastern end of the Niger delta. Topography of the area is shown on the right.

4.2.1 Methodology

Deltares' D-Flow modelling methodology using a flexible mesh for discretization of the modelled area is used here to check for increased possibility of flooding and inundation. D-Flow uses unstructured grids that can consist of triangles, pentagons and curvilinear grids and 1D channel networks all combined in one mesh. The unstructured grids (flexible meshes) can be modified to represent the complexity of a river and floodplain system. Calculations with D-Flow are based on 2D shallow water formulation and it uses the staggered grid convention; by which the water levels are defined at the calculation points and nodes and the discharges are calculated at the reaches. D-flow can be simulated in 1D, 2D or 3D.

4.2.1.1 Model setup

D-Flow requires boundary conditions at the upstream and downstream of the river reach, and the initial conditions of the system. For the modelling therefore, SRTM data was used to measure the river width, the heights of the tide gauges (Mean higher high water levels readings) between the upstream and the downstream points were used to determine the river slope, Manning's n values were used for river channel roughness. SLR values were added to tidal sea levels and used as the model downstream boundary condition. One month tidal data from two sources (UK Hydrographic office: Almiralty Easy Tide, 2017) and (Tides4Fishing, 2016) were compared to verify the range of water levels and the correlation coefficient was over 0.99 (figure 4.11). Initial conditions for the river is the MLLW (mean lower low water).

For river bathymetry, an unstructured grid was created for the Bonny River in D-Flow. The grid consists of curvilinear grids in the river channel and triangular grid an area with a small island where a tributary meets the New Calabar River; the two grid types were merged together to get a single grid (figure 4.11). D-Flow grids are made up of net nodes (corners of the cell), net links (cell edges that connect net nodes), flow nodes (cell circumference) and flow links (line connecting two flow nodes). In an ideal D-Flow grid topology, the cosine of the angle between the flow link and the net link (known as orthogonality), and the ratio of the areas of two adjacent cells (known as smoothness) should be equal to one (Deltares, 2016). The angle should therefore be 90^0 to ensure accurate hydrodynamic computations. The Bonny River has areas with sharp bends as well as distributaries, which make the creation of a perfect grid more difficult.

Figure 4.11. Unstructured grid created for Bonny - New Calabar River opened on Google earth image of the Niger delta. Yellow points show tide gauge locations.

The final grid used in this model therefore has orthogonality of 0.045 showing that the smallest angle between the net link and flow link is 87.4^0. In the model setup, the model upstream boundary condition used is set as zero because Bonny River flow is said to be primarily influenced by the tidal waters from downstream (Briggs, Okowa, & Ndimele, 2013; NDRMP, 2004a).

This was checked by running the model with constant discharge values, and the results with 250m^3/s and 120m^3/s showed there is little influence of river flow on the Bonny-New Calabar system (figure 4.12).

4.2.1.2 Model calibration

The model was run from 1st to 7th August, and the results were used to calibrate the model using measured tide gauge water level readings at observation point 8 on Bonny River. The model output was calibrated based on tidal water level measurements in the rivers, using measured roughness values and average flow velocity obtained from (Epete, 2012; Uzukwu, Leton, & Jamabo, 2014).

The setup was adjusted using Manning's n values to find the find the best model. The final calibrated model had a correlation coefficient of 0.97 with the measured data (figure 4.12). Data for observation point 8 (figure 4.14a) was used to calibrate the model; this is the location of a tidal gauge station that gives daily tidal water levels (Ford point in figure 4.11).

 The width of cross section 9 which is the nearest cross section to observation point 8, and the slope of the river were measured using SRTM DEM. The width, slope and maximum daily water levels were used to calculate the daily maximum discharge at that cross section using Manning's formula given as: $Q = \frac{1}{n} A R^{\frac{2}{3}} \left(\frac{\partial h}{\partial x}\right)^{1/2}$

(1)

Where Q=discharge, A= area of the cross section, R=hydraulic radius, n=Manning's roughness value, $\partial h/\partial x$ = slope. The calculated Q was also used as a calibration measure for the model.

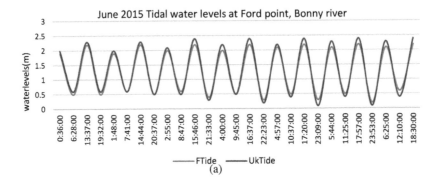

(a)

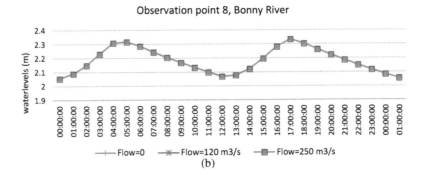

(b)

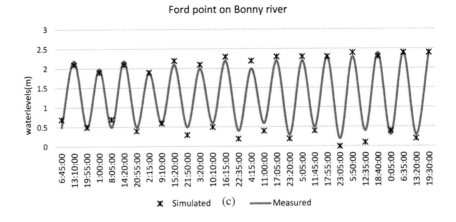

(c)

Figure 4.12. Comparison of two sources of tidal water level data for the Niger delta (a). Model results with Upstream flow values 250 m³/s and 120 m³/s. and 0 m³/s (b). Final calibrated model results compared with measured data at Ford point(c).

4.2.2   Model results

Water level /Water depth change

The downstream boundary was adjusted by adding different values of sea level rise from 0.2 to 1m. Wind speed has a great effect on coastal flooding as it contributes to storm surges and high wave heights. With high winds, sea water levels are elevated to orders of magnitude of the normal heights and push sea water further inland (NOAA, 2016). Wind velocity values[5] of 3.93m/s for the month of August are added to the model as it is important for coastal flooding.

The results show that channel water levels and water depths will be higher with SLR. Figure 4.13 shows change in water levels/depths at an observation point with SLR. This increase in water levels will flood inland areas at high tide (figure 4.14).

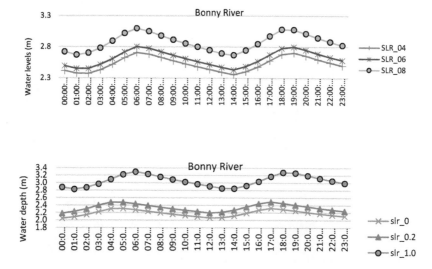

Figure 4.13.Increase in water levels (above) and water depths (below) with sea level rise, Bonny River. SLR_0 indicates zero SLR, SLR_0.4 indicates 0.4m SLR.

---

[5] Wind data measured by the Nigerian Meteorological Agency (NIMET) is obtained from (Adaramola et al., 2014).

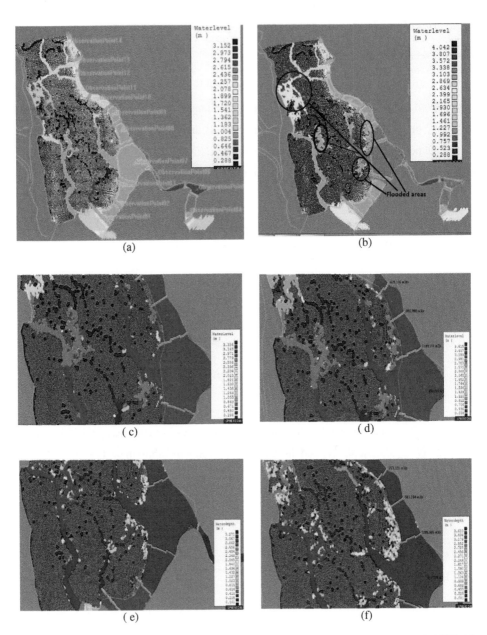

Figure 4.14. Water level change with sea level rise for the Bonny-New Calabar system. (a) No added SLR, (b) 80 cm SLR. Zoomed-in view of water levels of flooded areas with 60 cm SLR (c), 80 cm SLR (d). Water depths of flooded areas with SLR 60 cm (e), SLR = 1m (f).

Figures 4.14 (a) shows model results for water levels without SLR. The water is confined within the channels and the various creeks that crisscross the delta. With SLR however, model results show flooding of land areas (figures 4.14. b, c, d), with the flooded areas increasing with higher SLR values. Figures 4.14 (e) and (f) also show that water depths on flooded land areas are much higher than those of creeks; this implies permanent inundation of inland areas by SLR.

This simulation result show that some settlements located within the model area are at risk of flooding and inundation with high seawater under SLR. Settlements such as Timbikuku will be flooded with SLR values higher than 50cm, others like Eyamba, Okirika and Sara will be affected with SLR values as low as 20cm.

Velocity

The average flow velocity in the Bonny –New Calabar river system is 0.3m/s (NDRMP, 2004a; Epete, 2012; Uzukwu, Leton, & Jamabo, 2014), however with sea level rise, the river flow velocity will likely increase especially around bends and areas where the river cross section narrows; this is because the river system is dominated by tidal flow (Briggs, Okowa, & Ndimele, 2013). Figure 4.16 shows increase in sea water velocity at the sea boundaries as it enters the estuary. This implies fast flow of sea water, which can thus reach more areas upstream at high tides.

The results also shows faster flow of water on the floodplain area. This has the potential effect of changing the biodiversity by displacing bird and plant species that flourish in areas with slow moving/ almost stagnant water.

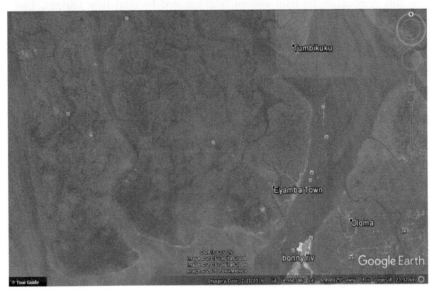

Figure 4.15. Google Earth images of study area showing some of the towns likely to be affected by SLR.

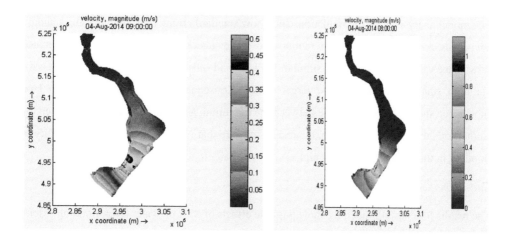

Figure 4.16. Velocity changes with rise in sea levels in Bonny River.

4.3 Conclusion

The modelling results covered in this chapter show that the lower Niger River will be affected by rise in sea levels. The effects include increase in flooding, increase in river water depths, inundation of low-lying areas and extension of flooding to areas further upstream (than would occur without SLR).

River modelling

For areas nearest to the coast, the results for the Nun River indicate that the effects of flooding from upstream will not be further exacerbated by SLR. However for river Forcados, there will be increase in the water depth of flooded areas. The increase in water depth depends on the amount of rise in sea levels which will be further exacerbated by land subsidence throughout the flooded areas.

The simulation results indicate that higher sea levels will cause flooding of more upstream areas. High sea levels downstream impede downward flow of flood waters which results in a backwater effect that floods more areas upstream. Consequently, although lateral flooding extent might not expand in the downstream areas, flooded areas will increase upstream.

For years with no flooding from upstream, SLR will cause coastal areas to start flooding earlier than usual. Moreover areas upstream of the Nun River which remain dry in normal years, will get flooded when sea levels rise.

The model analysis with 25 simulations show low standard errors, indicating that the standard deviations are within acceptable ranges. However, due to the large range of estimated subsidence values, the analyses of simulations with land subsidence show higher levels of error via high standard deviations and larger errors. Consequently, the simulations with land subsidence show a large difference between the maximum and minimum values. Although the amount of subsidence is uncertain, the results indicate that with subsidence, the probability of flooding in the Niger delta is very high.

Coastal modelling

The results for coastal modelling show possible effects of SLR on water level, flooding extent, water depth and the velocity of flow in Bonny River. SLR will cause acceleration of coastal flood waters and high tides inland, channel water depth and water levels will be much higher than at spring tide, and the extent of flooding will also be increased. Moreover change in model wetted area show possibility of inundation is high.

In this paper freely available satellite data are used to provide river bathymetry and river cross section width. The data is combined with freely available tidal data to calculate discharge at a tidal water level point. The data thus obtained was used as the model input for the Bonny-New Calabar river system. Use of unstructured flexible mesh for coastal flood/inundation model enables utilization of finer/coarser mesh resolutions at different parts of the modelling domain. In this study areas around bends where there is sudden change of river flow dynamics; thus information on velocity effects of SLR were shown and the complexity of the flow pattern was captured.

5

Vulnerability to sea level rise[6]

[6] This chapter is an edited version of the following publication:

Musa, Z.N, I. Popescu, & A. Mynett. 2014. "Niger delta's vulnerability to river floods due to sea level rise." *Natural Hazards and Earth System Science (NHESS)* 14: 3317-3329. Doi: 10.5194/nhess-14-3317-2014.

Vulnerability as a concept represents a potential damage and it is conditional upon the possibility of a hazard. Thus a system is said to be vulnerable when it has a high susceptibility to the effects of a hazard, and is unable to cope, recover or adapt (Balica, Popescu, Beevers, & Wright, 2013). System vulnerability assessment to a certain hazard gives a measure of the degree of damage that might likely occur if the hazard happens without mitigation/adaptation measures put in place. Vulnerability levels are varying within a system therefore indicators are used to determine and measure it. Such indicators can be ecological, political, technological and socio-economic factors of a system. The value of an indicator is used to represent the character of the system in a quantitative way (Cutter, et al., 2008). Consequently an assessment of vulnerability to SLR requires a method that takes into account various indicators that reflect the effects of the SLR on the vulnerability itself. Due to the complex nature of a coastal system, such methods include assumptions that simplify coastal processes in order to enable the assessments to be useful.

One method to determine the values of the indictors of vulnerability to river floods, due to SLR, is to represent data in Geographic Information Systems (GIS), which enables comparison and deduction on the relationships between the sources of the data. Heberger, et al., (2009) used GIS and hydrodynamic modelling to estimate the potential impacts of SLR on population, infrastructure, ecosystems and property, in case a major flooding event will occur on the river discharging into the sea. Data used for the assessment were: DEM's, base flood elevation data, population block data, hydrological data, tidal data, data on geology, built up area data, etc. The results combined inundation and erosion layers with population block layer to determine the population at risk. Similarly, a GIS based coastal vulnerability assessment was carried out by (Martin, Pires, & Cabral, 2012), based on physical and human induced vulnerability. The physical factors considered were: coastal systems, hydrology (sediment supply) and lithography while the human influence factors were road network, population density, population growth and urban land cover. The result was combined with an urban growth model to show the influence of anthropogenic factors on the final vulnerability of the area.

Another method for assessing vulnerability is the Coastal Vulnerability Index (CVI), which relates various factors that influence the degree of vulnerability of coastal areas in a quantifiable manner. The CVI concept introduced by Gornitz, et al., (1991) uses information about the coast to quantify the relative vulnerability of coastal segments to effects of SLR at a regional and national scale. In their study, Gornitz et al. (1991) assessed the vulnerability of the U.S coast to erosion and inundation effects of SLR by ranking sections of the coast according to their potential for change and relative importance for coastal management. Since 1991 the CVI methodology has been applied globally using different variables depending on the coastal area under study and the particular hazard being anticipated.

Pendelton, et al., (2010) and Dwarakish, et al., (2009) used six variables to assess the coastal vulnerability to sea level rise and coastal change for the northern Gulf of Mexico in Mexico and Udupi coastal zone in India, respectively. These six variables are geomorphology, coastal slope, mean wave height, mean tidal range, rate of shoreline change, and relative SLR, which are considered physical variables that characterise a coastal area, and relate to susceptibility of the shoreline to natural changes and its natural ability to adapt to changes in the environment. A similar methodology using different variables is demonstrated by Kumar & Kunte, (2012) for the Chennai East coast in India to calculate the possible areas of inundation due to future SLR and land loss to coastal erosion. Yin, et al., (2012), used elevation, SLR, slope, coastal geomorphology, shoreline erosion, land use, mean tidal range, and mean wave height to determine the areas of the Chinese coast that are most vulnerable to effects of SLR.

The CVI method is based on physical coastal variables and is therefore not easy to be used for coastal management; which would need variables related to social conditions and human impact on the environment in order to determine a good view on all aspects entailed by the vulnerability of coastal areas. Consequently modified CVI approach is developed, which includes variables that represent social, economic, and human-influence factors of the coast. Ozyurt & Ergin, (2009) propose an improved CVI for SLR, and apply the methodology to assess the impact of SLR for the Goksu Delta in Turkey. The approach uses seventeen physical and human influence variables, namely: rate of SLR, geomorphology, coastal slope, significant wave height, sediment budget, reduction of sediment supply, river flow regulation, engineered frontage, groundwater consumption, land use pattern, natural protection degradation, coastal protection structures, tidal range, proximity to coast, type of aquifer, hydraulic conductivity, depth to ground water level above sea level, river discharge, water depth at downstream. Result shows the vulnerability levels of defined coastal segments, to different types of impacts and indicates that human impact on the environment has the highest effect for inundation. The method however does not consider social variables. Mclaughlin & Cooper, (2010) include socio-economic variables in calculating a CVI for erosion in Northern Ireland. Their CVI included variables like population, cultural heritage, roads, railways, land-use and conservation status. The main outcome of their study is that socio-economic variables do not influence the scores of the CVI in a significant way. This result is due to the fact that socio-economic variables were assigned lower weights than to the physical variables. Indicator based studied such as the ones just cited use variables whose ability to change and respond to various effects of SLR (e.g. flooding) can be related to the systems susceptibility to the particular hazard under consideration. The results of such studies highlight the areas with characteristics that make them vulnerable to the effects of SLR; although the final proof of the vulnerability of an area will consist of several sources of information among which are numerical models and data obtained via field work.

The study presented herein uses the advantage of mapping CVI results in a GIS environment in order to analyse Niger Delta's physical, social and human influence on the environment in case that a flooding event will occur on the Niger River. The coastal vulnerability index obtained as such is a composite one, and it is called coastal vulnerability index due to SLR (CV$_{SLR}$I). In order to determine and analyse the CV$_{SLR}$I for the Niger Delta, seventeen variables that have relevance to coastal erosion, flooding/inundation and intrusion of sea salts (into underground water) are used (presented in Table 1). The option to use 17 variables is based on the data availability for these indicators as documented and suggested in studies by Gornitz, et al., (1991) for general coastal areas; and Ozyurt & Ergin (2009) for deltas. Similarly, Balica, et al., (2009) and Ozyurt & Ergin (2009) documented social and human influence variables that are important for determining the vulnerability of coastal areas. The variables are classified into exposure, susceptibility, and resilience classes based on their characteristics, following the methodology of Dinh, et al., (2012).

This chapter is structured in four parts. After the introduction and review of vulnerability methods above, a description of the applied methodology is given in section 2. Results are presented in section 3, followed by conclusions in section 4.

5.1 Vulnerability assessment methodology

Gornitz (1991) defined CVI on n number of physical ranked variables (x_1.....x_n), as:

$$CVI = \sqrt{\frac{x_1 \cdot x_2 \ldots \ldots x_n}{n}} \tag{1}$$

In equation (1) n represents the number of ranked variables.

According to the CVI method, local variable values are measured and/or analysed and compared with documented ranges of values for that variable. The comparison allows a ranking of physical variables that shows the level of vulnerability.

Variables can be categorised in classes of exposure, susceptibility and resilience. Dinh et al. (2012) defined a coastal vulnerability index based on exposure, susceptibility and resilience factors as:

$$CVI = \frac{E * S}{R} \tag{2}$$

Where E are exposure factors, S susceptibility factors and R resilience factors.

The exposure variables are those inherent qualities of the system that position it for a likely hazard impact; they describe what is exposed to the threat (Cutter, et al., 2008). Susceptibility variables are the characteristics of the exposed system that influence the level of harm from hazards (Birkmann, 2007). The resilience of a system implies the ability to adapt and even utilize the disaster as an opportunity for the future; thus resilience variables enable a system to cope and reduce the possible impact of the disaster on the exposed population.

While equation (1) enables the simplified combination of variable rankings to calculate the CVI for exposure, susceptibility and resilience, equation (2) enables the combination of the three indices to allow a ranking of vulnerability that acknowledges the importance of systems resilience. Exposure and susceptibility variables increase the vulnerability of systems, while resilience variables enable systems to withstand and reduce the vulnerability to hazards. Consequently, the methodology used in the present research combines the two methods into a composite index which multiplies the exposure index by the susceptibility index and divides the product by the resilience index. Because CVI can refer to different regions and causes, further on, the index of vulnerability to river floods in coastal areas due to SLR is referred as Coastal Vulnerability to SLR Index ($CV_{SLR}I$). The proposed methodology to evaluate $CV_{SLR}I$, has the following application steps:

1. choose variables that are relevant to the coastal processes in the study region;
2. classify variables in exposure, susceptibility and resilience;
3. define coastal segments and determine for each of them the values of the variables chosen in the first step;
4. use equation (1) to calculate the CVI for exposure, susceptibility and resilience elements (e.g. CV_EI, CV_SI, and CV_RI respectively);
5. use equation (2) to compute the $CV_{SLR}I$ for each defined coastal segment, i.e.:

$$CV_{SLR}I = \frac{CV_EI * CV_S I}{CV_R I} \tag{3}$$

6. compare the $CV_{SLR}I$ with results obtained for CVI based on physical variables only;
7. indicate (through the $CV_{SLR}I$) the coastal segments that are most in need of intervention in response to socio-economic conditions

The developed methodology is herein exemplified on the case of the Niger delta, however its applicability is valid to any coastal area.

Rankings and ranges of the variables in table 4 are not the same across different systems, but depend on the measured values. Three to five classes of ranking are found in the literature Kumar et al (2010), and Kumar and Kunte (2012) use three classes (i.e. low, medium, high); Yin J. et al, (2010), use four (low, medium, high, very high); while Dinh et al, (2012), Pendelton et al (2010), Ozyurt & Ergin (2009), Thieler & Hammer-Kloss, (1999) and Gornitz (1991), rank the measured ranges into five classes from very low to very high. The later approach is used in the present study of the Niger delta, considering that such a refined classification will reduce considerably the uncertainty in computation of vulnerability. Table 2, 6 and 4 shows the ranges of values of exposure, susceptibility and resilience variables respectively, as considered in present research, as well as their ranking from 1 (very low) to 5(very high).

Table 4. List of selected variables for vulnerability assessment

Variable (class)	Data type	Data Source
Topography (E)	SRTM DEM	srtm.csi.cgiar.org/
Coastal slope (E)	SRTM DEM	Validation map from NASRDA data archives; srtm.csi.cgiar.org/
Geomorphology (E)	Geomorphologic map of Nigerian coast	www.niomr.org
Relative SLR rate (E)	Relative sea level rise rates for Niger Delta Atlantic coast	www.niomr.org
Annual shoreline erosion rate (E)	Measured annual erosion rate for the Nigerian coast	www.niomr.org
Mean tide range (E)	Tidal data for Nigerian coast	www.niomr.org; www.wXtide32.com
Mean wave height (E)	Wave height data for the Nigerian coast	www.niomr.org
Population density (E)	Population distribution data per local Government area.	Nigerian National Population commission. www.population.gov.ng
Proximity to coast (E)	NigeriaSatX imagery and settlement map of Niger Delta.	NASRDA data archive
Type of aquifer (S)	Data on aquifer types in the Niger delta	Niger Delta Regional Master Plan (NDRMP)- Environment and Hydrology report
Hydraulic conductivity (S)	Data on aquifer properties in the Niger delta	NDRMP- Environment and Hydrology report

Reduction in Sediment Supply (S)	Estimate on reduction in sediment supply from the Niger River.	NDRMP- Environment and Hydrology report
Population growth rate (S)	Inter-census data	www.population.gov.ng
Groundwater consumption (S)	Data on % ground water consumption	NDRMP- Environment and Hydrology report
Emergency services (R)	Information about presence and type of emergency services	National Emergency management Agency, (NEMA) www.nemanigeria.com
Communication penetration (R)	Data on settlement type, size, and location.	NASRDA data archives.
Availability of shelters (R)	Info on provision of shelters	NEMA, www.nemanigeria.com

Note: E =EXPOSURE, S= SUSCEPTIBILITY, R= RESILIENCE
P= PHYSICAL, SO = SOCIAL, HI= HUMAN INFLUENCE

The applied methodology divides the coast in segments. For each segment a $CV_{SLR}I$ is calculated, however these values present a wide range between a minimum (min) and a maximum (max) value, therefore results are normalized between 0 and 1 using the relation:

$$NV = \frac{value - min}{max - min} \qquad (4)$$

Where NV is the normalised value of the variable; *value* is the calculated index value for a coastal segment; *max* is the maximum value in that index; and *min* is the minimum value in that index.

The selected indicators used for determining the CV$_{SLR}$I are detailed further.

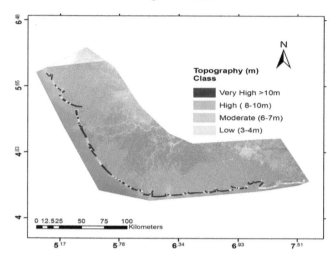

Figure 5.1. Niger delta topography classification

5.2 Selected indicators for Exposure

The exposure indicators are selected based on their influence on coastal flooding, inundation, sea water intrusion to ground water sources and coastal erosion. All the chosen variables are physical properties of the coast except 'Proximity to Coast' which is a human related variable.

5.2.1 Topography

The topography (elevation) of an area above the mean sea level influences how much of it will be impacted by rising sea levels because low lying areas offer less resistance to inundation in times of flooding and storm surges (Van, et al., 2012). The elevation of the Niger Delta is extracted from SRTM DEM data using ERDAS Imagine 9.1 topographic analysis tools. The coastline has an average elevation between 0 and 10m above sea level, which is ranked as defined in table 5. The coastline topography mapping, based on the defined ranking in table 5, is shown in Figure 5.1. It can be noticed that the eastern end (from Bonny) has 'medium to high' topography (3-7m a.m.s.l), which makes the delta susceptible to flooding due to river flow and to storm surges coming from the sea.

5.2.2 Coastal Slope

The slope of a coastal area is the degree of steepness with reference to the surrounding land. Slope determines the minimum level of water that can penetrate and inundate an area; therefore areas with lower or gentler slopes are more vulnerable to waves and tide action than areas with steeper slopes (Aich, et al., 2014). The delineation and classification of the coastline slope ranges between 0% and 2.5%. Figure 5.2 shows the classification of the slope and the fact that the eastern end (from Bonny) has a slope of 0.1%-1%, which gives it a 'high' to 'very high' vulnerability ranking; making it highly susceptible to inundation.

5.2.3 Geomorphology

Geomorphology describes landforms and processes that lead to the formation of landform patterns. The type of landform found on the coast determines its degree of vulnerability to erosion and its level of resistance to wave forces. Vulnerability ranking based on geomorphology is done such that cliffs and rocky areas have low vulnerability; lagoons and estuaries have high vulnerability; while beaches, deltas, and barrier islands have very high vulnerability (Pendelton et al, 2010). The Niger Delta geomorphologic zone is characterized by deltaic, sandy beach, and estuarine landforms. These characteristics (see table 5) gives it a 'high' to 'very high' ranking and makes it very susceptible to erosion and wave action.

5.2.4 Relative Sea Level Rise

Relative sea level/annum at the local level is a measure of the height of the sea above a certain datum averaged over a year and measured using tide gauges (Yin et al., 2012). The higher the sea level rise rate the more vulnerable an area is compared with those with lower rates of rise in sea levels. Satellite altimetry measurements (1993 to 2010) over the Niger delta coast show eustatic sea level rise rates of 3.03-3.39mm/year (Rosmorduc, 2012). In addition the Niger delta is subsiding at a rate of 25-125mm/year, which classifies it as a 'very high' relative SLR (see table 5).

Table 5. Data range and ranking of the exposure CV$_{SLR}$I variables

	Variables	Class	**Ranking of values**				
			Very low (1)	Low (2)	Moderate (3)	High (4)	Very high (5)
1.	Topography	P	>10m	8-10m	6-7m	3-5m	0-2m
2.	Coastal Slope	P	>3% - 4%	2% -3%	1% -2%	0.5% -1%	0.1% - 0.5%
3.	Geo-morphology	P	Rocks	Cliffs	Vegetated coasts	Lagoons estuaries	Barrier islands, beaches, deltas
4.	Relative SLR rate	P	0 - 1mm	1 -2mm	2 -3mm	3-4mm	>4mm
5.	Annual shoreline erosion rate	P	0-1m	1-5m	5-10m	10-15m	>15m
6.	Mean tidal range	P	>6m	4-6m	2-4m	1-2m	<1m
7.	Mean wave height	P	0.3- 0.5m	0.5- 0.8.m	0.8 -1.1m	1.1 -1.4m	>1.4m
8.	Population density	SO	<100 people/km^2	100-300 people/km^2	300-500 people/km^2	500-800 people/km^2	>800 people/km^2
9.	Proximity to coast	HI	>800m	600-800m	400-600m	200-400m	100-200m

Note: *P= PHYSICAL, SO = SOCIAL, HI= HUMAN INFLUENCE*

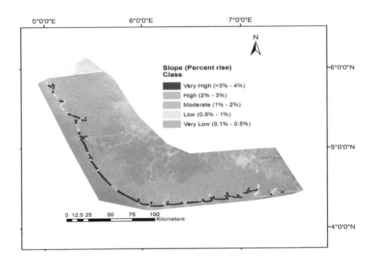

Figure 5.2. Niger delta slope classification

5.2.5 *Annual Shoreline Erosion Rate*

The degree of erosion of a coastal area influences its response to rising sea levels. In view of coastal vulnerability, areas that are undergoing erosion will have high vulnerability while areas of accreting sediment will have low vulnerability (Kumar and Kunte, 2012). Niger delta values for annual erosion as published by NIOMR (2010) are: Escravos 20-25m/year, Forcados 16-20m/year, Brass 15-20m/year, and Bonny 10- 14m/year. These are the values considered in the present study, because they cover the Niger Delta from west to east. The values show that the Niger delta has a 'high' to 'very high' ranking (Table 5) and is therefore very susceptible to more erosion from SLR.

5.2.6 *Mean Tidal Range*

The tidal range gives the difference between high and low tides and is linked to permanent and episodic hazards from sea level rise and storm surge (Yin et al., 2012). In view of coastal vulnerability, areas with large tidal ranges have higher vulnerability than those with lower ranges. Mean tidal range is in general determined based on long-term tidal data. In case such data is not available, hydrodynamic models are used to predict tidal levels based on tidal stations located within the areas of interest (Kumar and Kunte, 2012). Values of the tidal range for the Nigerian coast are generated using the wXTide32 tidal model, which predicts tides based on the algorithm developed by the U. S. National Oceanic Service. Niger Delta measurements from eight tidal stations, along the delta coast, are used in the model. The results show a gradual increase from 1.74m in the west, around Forcados River to 2.57m in the east at Bonny River. The range (1.74- 2.57m) has 'moderate' to 'high' ranking (see Table 5); therefore the Niger delta is susceptible to storm surge and sea level rise.

5.2.7 Mean Wave Height

Waves move coastal sediments from one place to another. The linear wave theory gives the wave energy as:

$$E = \frac{1}{16}\rho g H^2$$ (5)

Where E= energy and H = wave height.

According to equation (5) wave energy is directly proportional to the square of wave height, therefore the wave height can be used as a proxy for wave energy (Yin et al; 2012). Areas with high waves are more vulnerable than areas with low wave heights as they have more energy to move materials offshore. Values obtained from NIOMR (2010) give wave heights of 1.5m for the western to middle Niger delta (from Jalla to areas around Okumbiri), and 0.5-1.5 m for the eastern end. These values have a 'high' to 'very high' ranking (see Table 5) and make the coast susceptible to flooding, erosion, storm surge and inundation.

5.2.8 Population Density

Areas with high population density have a higher vulnerability than those with lower population density (Mclaughlin, Mckenna, & Cooper, Socio-economic Data in Coastal Vulnerability Indices: Constraints and Opportunities., 2002). The presence of human settlement increases the value of risk, the likelihood of erosion and modification of the coastal area. The Niger delta population distribution data, as given by the local Government area shows that many settlements in the eastern end (from Bonny) have higher than 500 people per km^2; hence there is 'high' to 'very high' vulnerability risk to SLR (see Table 5).

5.2.9 Proximity to Coast

The proximity of a settlement, infrastructure or land to the coast determines the level of its exposure to the effects of sea level rise such as storm surges, floods, erosion and wave action. Present study considered distances from shore using the 2012 NigeriaSatX satellite imagery. Locations of settlements within 0-1500m of the coastline were determined and ranked. Table 5 shows that the shorter the distance from the coastline, the more vulnerable the settlement is to the effects of SLR.

5.3 Selected indicators for Susceptibility and Resilience

Based on the available data and the influence of social and human factors on the extent of damage that could occur at the occurrence of SLR, eight indicators of susceptibility and resilience are selected in the present study. The resilience indicators are the social variables that increase the ability of victims to cope with floods, inundation, loss of land from erosion, and intrusion of sea salts.

Like the exposure variables, the documented range of values/characteristics of the susceptibility and resilience variables are ranked from 'very low 'to 'very high' vulnerability, as detailed in Table 6 and Table 7, respectively. The indicators are explained in detail below.

5.3.1 Type of Aquifer

The type of aquifer in a given area determines how vulnerable the ground water is to salt water intrusion. Confined aquifers are overlain by materials with poor permeability and are therefore less vulnerable to contamination than semi confined and unconfined aquifers, which allow interaction with the surface. Data from the NDRMP (2004a) was used to rank the Niger delta coastal aquifers and the results shows that the coastal aquifers are unconfined. Table 6 shows that unconfined aquifers have 'very high' vulnerability; therefore the Niger Delta is vulnerable to salt water intrusion from SLR.

5.3.2 Aquifer Hydraulic Conductivity

Hydraulic conductivity of an aquifer is the ability of the aquifer to transmit water. Its value depends on the type of material, for example gravel/sand have a higher hydraulic conductivity than clay/silt and therefore transmit water more easily. Areas with high hydraulic conductivity are more vulnerable to the effects of SLR than those with low hydraulic conductivity (Ozyurt & Ergin,) Improving Coastal Vulnerability Assessments to Sea-Level Rise: A New Indicator

Based Methodology for Decision Makers., 2010). The hydraulic conductivity of coastal aquifers in the Niger Delta ranges from 0.0002m/d to 120.6m/d (NDRMP, 2004a). Coastal segments with hydraulic conductivities higher than 41m/day have a 'high to very high' vulnerability ranking (Table 6) and are vulnerable to salt intrusion from SLR.

5.3.3 Reduction in Sediment

Building of dams and other control infrastructure in the upstream of coasts impede the flow of sediments and reduce the natural nourishment of delta areas (IPCC, 2007b). Areas where the percentage of sediment reaching the coasts is sustained over long period of time have less vulnerability compared to areas where only a percentage of the normal sediments reaches them (Ozyurt and Ergin, 2010). The sediment supply to the Niger delta is 70% less than in the past, due to construction of dams in the upstream (NDRMP, 2004a). The value (i.e. 70%) for reduction in sediment supply gives a 'very high' vulnerability (Table 6), which makes the Niger delta susceptible to erosion from SLR.

Table 6.Data range and ranking of the susceptibility $CV_{SLR}I$ variables.

	Variables	Class	Ranking of values				
			Very low (1)	Low (2)	Moderate (3)	High (4)	Very high (5)
10.	Type of aquifer	P	Confined		Leaky confined		Unconfined
11	Aquifer hydraulic conductivity	P	0-12m/d	12--28m/d	28-41m/d	41-81m/d	>81m/d
12	Reduction in sediment supply	HI	30%	40%	50%	60%	70%
13	Population growth rate	SO	0%	<1%	1-2%	2-3%	>3%
14	Groundwater consumption	SO,HI	<20%	20-30%	30-40%	40-50%	>50%

Note: *P= PHYSICAL, SO = SOCIAL, HI= HUMAN INFLUENCE*

5.3.4 Population Growth Rate

Population growth affects the environment in various ways with highly populated areas facing greater environmental challenges (UNFPA, 2008). High population growth rate will increase the number of people likely to be affected by the effects of SLR therefore areas with lower growth rate will have less vulnerability compared with those with higher growth rate. Inter-census data of the Niger delta (1991 to 2006) shows a growth rate of 2.9-3.1% which gives a 'high to 'very high' vulnerability (Table 6).

5.3.5 Ground Water Consumption

Inland intrusion of sea salts is likely to pollute underground aquifers and cause shortage of drinking water in coastal areas. Areas that depend on ground water as the main source of drinking water are more vulnerable than those with low dependence on ground water. Data on groundwater consumption in the Niger delta, as compiled by NDRMP (2004a) shows the percentage of households/settlement that depend on groundwater sources (boreholes and wells) for drinking and domestic use. Some areas have over 40% dependence on groundwater giving them a high ranking. People living in such areas are vulnerable to salt water intrusion due to SLR.

Table 7.Data range and ranking of the resilience CVSLRI variables.

| | Variable | Class | **Ranking of values** | | | | |
			Very low (1)	**Low (2)**	**Moderate (3)**	**High (4)**	**Very high (5)**
15	Emergency services	R, SO	Absent No settlements on segment	Only present at the state level. (Remote area)	Only present at the local government level. (Village community)	Present in community but not formally trained. (Settlement located far from HQ).	Present in the community and formally trained. (City/Local government HQ/ LG oil company).

| 16 | Communica tion penetration | R, SO | None | Only through direct contact; Access to radio communi cation; Remote areas | Traditional rulers/ town criers. Access to radio communicati on. Village settlement | Access to print and electronic media. Town/ settlement located close to oil company. | Print and electronic media. City/ Proximity to Local Government headquarters. |
| 17 | Availability of shelters | R, SO | Absent | Available but not equipped with relevant facilities | Available/eq uipped with relevant facilities but located in another community | Available/e quipped with relevant facilities but only accessible by boat | Available/equip ped with relevant facilities and accessible by road/boat |

Note: *P= PHYSICAL, SO = SOCIAL, HI= HUMAN INFLUENCE*

5.3.6 Emergency Services

Emergency service personnel are usually trained in first aid and search-rescue operations to enable them combat consequences of disasters. In rural remote communities these trained personnel are not available at the onset of disasters. Communities with trained and equipped emergency services are more resilient to the impacts of SLR compared to those without. In Nigeria, emergency services at the local level are coordinated by the Local Emergency Management Agency (LEMA) which establishes trained local community structures made up of local associations, religious bodies, clubs, schools etc. (NEMA, 2010a). Due to the presence of LEMA in every local government area in Nigeria, present study assumes that local community structures exist in all the Niger delta communities. However, the Niger Delta coast has small and isolated fishing communities which are less likely to have schools. The resilience ranking for such isolated communities is 'very low' (see Table 7).

5.3.7 *Communication Penetration*

The channel of communication determines the number of people to whom information reaches as well as the quality of information provided. In Nigeria, NEMA through its disaster prevention strategy provides information about impending disasters to vulnerable communities via print and electronic media as well as informal channels like traditional rulers, religious leaders, etc. (NEMA, 2010b). NEMA (LEMA) staff who disseminate this information are found in the Local government headquarters. Many settlements in the Niger delta are located far away from the local government headquarters and might not be easily reached. People living in such remote areas have less access to quality communication and are therefore less resilient to the effects of SLR, as compared with those living in cities (Table 7).

5.3.8 *Availability of shelters*

During a disaster, people are evacuated to shelters administered by trained personnel. Access to shelters determines the number of people that can be rescued in good time and helps restore later on the affected community (NEMA, 2010c). Areas with buildings located on safe sites that can be used as shelters are more resilient to the impacts of SLR than those without. In Nigeria buildings located on unaffected sites are used as shelters during flooding (e.g. schools), but where none is available emergency shelters are erected. The elevation of the Niger delta is generally low as shown in Figure 5.1 therefore in the events of flooding, evacuation camps have to be erected. This gives the Niger delta a 'very low' resilience ranking (Table 7).

5.4 *Results and discussion*

In order to calculate the $CV_{SLR}I$ for the 450km of the Niger delta coast, 54 coastal segments are considered (figure 5.3). The segment division is based primarily on three main elements; elevation (figure 5.1); change in slope (figure 5.2); and the presence of large estuaries. Elevation and slope are important factors for flooding since elevation determines the lowest level of water that could flood an area, and slope affects the flooding extent over an area. Therefore each of the 54 segments shown in figure 5.5 defines an area whose characteristics (slope and/or topography) makes it different from the neighbouring segments. Sizes of the segments differ from one another in length, however on average, the segment width is 4km inland.

For each coastal segment, the exposure, susceptibility and resilience indicators are calculated and ranked. The range of results for the Niger delta coastal segments are normalized using equation 4 and classified into five vulnerability classes (very low, low, moderate, high and very high) based on percentile ranges. Accordingly, the calculated results give the following ranges of vulnerability: 0.0-0.08 ('very low'), 0.08-0.19 ('low'), 0.19-0.26 ('medium'), 0.26-0.35 ('high'), and 0.35-1.0 ('very high').

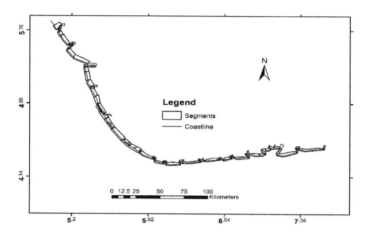

Figure 5.3. The 54 Niger Delta's coastal segments assessed for vulnerability to SLR.

As an example of the indicator ranking for the Niger delta coast, segments 1-4, 52 and 54 are presented in table 8. The most vulnerable segment (number 52) has a low slope (<1%), low topography (3-5m), estuaries, very high hydraulic conductivity (>81m/day), very high population density (>800 people/km^2), and settlements within 100-200m of the coast. These attributes have thus made it highly vulnerable to SLR. On the other hand, the least vulnerable segment, number 1, has a high slope (>4%), a topography higher than 10m, is uninhabited with no coastal infrastructure, and a very low hydraulic conductivity (0-12m/day). These attributes give it a very low vulnerability to SLR. (Segment count is from left to right).

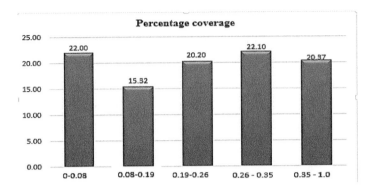

Figure 5.4. SLR CVI values over the Niger delta coastline

Figure 5.4 shows a plot of the calculated $CV_{SLR}I$ for the Niger delta coastal segments. Analysing the results it is seen that, 37.3% of the coastline has 'very low' to 'low' vulnerability, 20.2 % has 'moderate' vulnerability, while 42.5% have 'high' to 'very high' vulnerability; which is shown in figure 5.5.

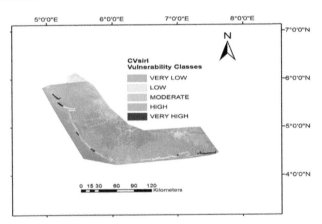

Figure 5.5. Niger delta coastal vulnerability levels, calculated using the CVSLRI. Areas with medium to very high vulnerability will need protection

In figure 5.5 the eastern end of the Niger delta (from Bonny to the southern end of Opobo (made up of six coastal segments: 49-54) is the longest stretch with very high vulnerability to SLR. As shown in the case of segment 52, such areas with 'high' to 'very high' vulnerability are characterized by 'very low' to 'low' slopes, 'very low' to 'low' topography, 'high' to 'very high' mean wave heights, unconfined aquifers, presence of coastal infrastructure and 'high' population density, etc. These variables represent physical coastal properties, human influence, and social properties. The presence of human influence variables such as coastal infrastructure and high population density, increase the probability of damage to lives and property when a disaster occurs. The combination of these properties has made the coastal segments highly vulnerable to SLR. The coastal segments classified as highly vulnerable to SLR will require mitigation measures to be applied against SLR.

The advantages of using a method such as the $CV_{SLR}I$ include the fact that it takes into account existing social structures (in terms of favourable places to live/ invest in infrastructure) and shows the level of vulnerability of choice areas.

Table 8. Ranking per indicator and CVI results for six segments

Segment no	1	2	3	4	52	54
Variable (factor)	**Ranking**					
Topography (e)	1	1	2	1	4	2
Coastal slope (e)	1	1	4	4	4	5
Geomorphology (e)	5	4	5	5	4	4
Relative sea-level rise rate (e)	5	5	5	5	5	5
Annual shoreline erosion rate (e)	5	5	5	5	4	4
Mean tide range (e)	4	4	4	4	3	3
Mean wave height (e)	5	5	5	5	3	3
Population density(e)	1	4	4	4	5	5
Hydraulic conductivity (s)	1	1	4	4	5	2
Proximity to coast(e)	1	4	1	4	5	5
Reduction in sediment supply (s)	5	5	5	5	5	5

Type of aquifer (s)	5	5	5	5	5	5
Population growth (s)	5	5	5	5	5	5
Groundwater consumption (s)	2	2	2	2	3	3
Emergency services (r)	1	1	3	3	1	5
Communication penetration (r)	1	1	4	4	1	5
Shelters (r)	1	1	1	1	1	1
$CVI_{exposure}$	16.68	59.64	94.29	133.32	180	141.42
$CVI_{susceptibility}$	7.06	7.06	14.13	14.13	19.36	8.65
$CVI_{resilience}$	0.57	0.57	1.99	1.99	0.57	2.88
$CVI_{slr} = \frac{CV_E I \cdot CV_S I}{CV_R I}$	206.6	738.7	669.6	946.6	6113.68	424.75
Normalized result	**0.06**	**0.35**	**0.32**	**0.4**	**1.0**	**0.28**

For example the ranking of segments 1 and 2 for physical variables 1 to7 (table 8) are similar, however their vulnerabilities are very different (table 8), since in the $CV_{SLR}I$ method human influence variables differentiate between the vulnerabilities of the two segments. While segment 1 has a 'very low' vulnerability, segment 2 has a 'high' vulnerability due to its high population density and presence of many settlements along the coast. If the CVI calculation was based on physical factors only, both segments will have similar vulnerability and segment 2 will be given a 'very low' vulnerability ranking, consequently, it will not be included in any adaptation plan. Thus $CV_{SLR}I$ results differentiate the levels of intervention needed on coastal segments that might have the same physical properties but different social conditions. Another advantage of the $CV_{SLR}I$ is that it includes, in the vulnerability assessment, the human modifications of the coastal environment. Human influences (e.g. construction of sea walls, groins, ports) add to the overall cost of impacts of coastal hazards, therefore there is need to capture them in a vulnerability assessment. Moreover, $CV_{SLR}I$ ranking of vulnerability acknowledges the importance of systems resilience in reducing the potential effects of SLR. The $CV_{SLR}I$ however, requires a wide range of data collection for the physical, social, and human influence factors which might not be readily available. Different variables might be available in countries within the same region, making comparison difficult.

5.5 Conclusion

Highly vulnerable coastlines expose the inland areas to effects of SLR, serving as a gateway for inundation, storm surge and coastal erosion. The results of the $CV_{SLR}I$ for the Niger delta shows that 42.6% of the coast is highly vulnerable to effects of SLR like flooding, erosion, and salt water intrusion into underground aquifers. These areas of the coast need to be protected against the negative effects of SLR.

Human influence on coastal environments can affect sediment supply and accelerate erosion, and should therefore be captured in vulnerability assessments. Analysis of social and human influence variables show that in terms of type of aquifer, aquifer hydraulic conductivity, population growth, sediment supply, groundwater consumption, the Niger delta is vulnerable to the effects of SLR. Moreover the location of many settlements in remote areas, far away from the local government headquarters, reduces the value of resilience to the effects of SLR.

Studies such as the one presented herein serve as an input for taking mitigation measures and helping decision makers to assess the effects of their measures in the function of the river system under consideration (Jonoski & Popescu, 2012; Popescu, Cioaca, Pan, Jonoskia, & Hanganu, 2015); The results of this study can provide a complementary source of data for the decision makers in planning mitigation/adaptation strategies for the Niger delta. For example, the map (in Figure 5.5) can be used alongside other data to identify those areas that are most likely to be affected by flooding from the Niger River before a flood occurs. With the Niger River flooding frequently in recent times, mitigation/ adaptation strategies can be planned for vulnerable areas and not for the entire geographical region all at once. Also, the evaluation of resilience for the coastal segments which shows the ability of the system (people) to cope and adapt to the disaster can aid in mitigation/ adaptation planning. Under resilience, we evaluated three variables: emergency services, communication penetration, and availability of shelters; which are services directly provided by the decision maker. The evaluation results can be useful in channelling more services to areas most in need.

Global studies undertaken by Ericson, et al., (2006) and Nicolls & Mimura, (1998) rank the entire Niger delta as having moderate vulnerability. Such ranking has been used in literature as the 'condition' for the Niger Delta, even though it was only based on population likely to be displaced. The evaluation presented herein shows that parts of the Niger delta are highly vulnerable to SLR. The results can be used to identify focus areas that need modelling of flooding to aid mitigation and adaptation planning. Hence, a combination of this study results with physical models of flooding in the Niger delta will provide a much better picture of the effects of sea level rise for the decision makers.

The segment division used in the study has constrained the scale of CVI calculations and reduced the possibility of generalizing variable values along the Niger delta coastline. For example in the case of the variable 'population density', since the segments divide the coastline into smaller areas we were able to use data provided per local government area to classify vulnerability instead of data per state (which is a much larger scale). Present study is however limited to onshore areas and does not include the vulnerability of offshore areas or mitigation/adaption to SLR options. The mapping of vulnerability as presented in the study is within limited bounds of the data accuracy and the scale of the study. Even though the local data used is acquired from official sources (see Table 4), there might still be uncertainties in the data collection and methods of processing that cannot be accounted for, because officially data is accepted as reliable by the authorities in charge of managing the delta.

The influence of scale is such that some of the variables used in the study presented herein might not be applicable at a larger scale; for example in a study of the vulnerability of the entire West African coast, there will be several rivers to be taken into account in measuring the variable 'reduction in sediment supply', or the variable 'population growth rate' (which is difficult to be included because several countries in the region have different data and measurement techniques of population growth).

Such a study however would complement the overview of decision makers of the vulnerability in the area, and will allow them to take adaptation measures that would address in a coherent manner both the Niger Delta as well as the Nigerian Coastline.

6

Resilience to sea level rise[7]

---

[7] This chapter is an edited version of the publication:

Musa, Z. N., Popescu, I. & Mynett, A., 2016. Assessing the sustainability of local resilience practices against sea level rise impacts on the lower Niger delta. Ocean & Coastal Management 130 (2016) 221-228. http://dx.doi.org/10.1016/j.proeng.2016.07.566

In order to effectively reduce the effects of a hazard via adaptation, it is important to assess the existing resilience of the system it will affect (Brooks N. , 2003). Studying the existing resilience level of a system shows how to adjust its characteristics that have direct effects on the resilience (e.g. topography, slope, population density, access to clean water). The resilience of a system implies the ability to adapt and even utilize the disaster as an opportunity for the future (Alwang, Siegel, & Jorgensen, 2001). Just like vulnerability, resilience is a potential state that exists within a system which enables it to cope after it encounters a hazardous event. Components of resilience can be physical (e.g. sea walls), or non-physical (e.g. past experience).

Resilience strategies can be introduced formally into a system, or happen through informal sequences of events. Formal strategies are usually planned, documented, and budgeted for before implementation; thus they have 'needs' and feasibility studies undertaken before implementation. On the other hand, informal resilience strategies are traditions and knowledge/expertise developed by local people over many years. Informal strategies include emergency coping mechanisms acquired during hazards; efforts by individuals and communities to save their lives and properties. Informal interventions are mostly reactive in nature (Bierbaum, et al., 2013), depend on the available individual resources, and are implemented on a small scale.

Formal strategies are usually part of the government's deliberate efforts to improve the land and livelihood of the people (Bachmair, et al., 2012). The Nigerian government at the state and federal levels have embarked on projects that will mitigate flooding and erosion. For erosion control, sand nourishment, revetment, sand replenishment, sheet pile and revetment, land Reclamation and sheet Pile works, earth dykes, and gully reclamation works have been implemented at different sites in the Niger delta (NDRMP, 2004a). The NDDC was able to undertake 47 shoreline protection projects and 90 water supply projects between 2001 and 2003 (NDRMP, Niger Delta Regional Master Plan. Chapter 2: Regional developement efforts, 2004c). For river floods and sea level rise control, bridges, canals and channels have been constructed as flood control measures (NEST, 2011; NDDC, 2014). Also, in order to address the problem of inadequate drinking water supply in the Niger delta, the government sinks boreholes.

Some formal interventions are undertaken by non-governmental or international bodies but regulated and documented by government. Such organisations present in the Niger delta include: United Nations agencies, the World Bank, representatives of international development initiatives from developed countries and donor agencies (Bachmair, et al., 2012). These groups of Non-Governmental-Organisations (NGO's) sponsor environmental, gender and vulnerability support projects that are relevant to the sustainability of the area, and play a role in interventions against possible effects of sea level rise. For example, the Building Nigeria Response to Climate Change (BNRCC) group sponsored by the Canadian International Development Agency (CIDA), taught locals in Calabar about rain water harvesting (to reduce reliance on groundwater), flood water diversion, sand bagging (against flood), and planting of bamboos (to reduce erosion), (NEST & Woodley, Reports of Pilot Projects in Community-based Adaptation - Climate Change in Nigeria. Building Nigeria's Response to Climate Change (BNRCC), 2011). Such efforts help to sensitize and train locals thereby increasing their resilience to hazards.

These efforts by the government, multinational agencies and NGO's have provided help in various forms, however only a limited part of the Niger delta have been reached. Consequently in many parts of the delta, the local people have to do major constructions like embankments themselves (NEST, 2011). This chapter presents the methods used by the local people for adaptation and mitigation of flooding, erosion and inundation in Niger delta. As inundation can be a permanent condition for the locals to deal with, an analysis of characteristics of the Niger delta and its vulnerability to inundation under SLR conditions are undertaken using GIS. The chapter aims to review the current situation and show the possible future inundation conditions in order to encourage inclusion of local resilience practices in future adaptation planning for the Niger delta.

6.1 Methodology

The methodology involves a review of local practices used in the Niger delta to combat flooding, erosion, inundation, and intrusion of sea salts. Local responses to these natural hazards and the degree of effectiveness of the methods in enabling the people return to their normal lives are reviewed. The data used include published papers, reports, and analysis on adaptation efforts in the Niger delta.

GIS is used to map areas vulnerable to inundation by 2030 and 2050 using the bathtub approach. Storm surges, large wind driven waves and high tides can raise sea water levels far above MHHW levels and increase likelihood of flooding and inundation of coastal areas

(NOAA, 2016). The modified bathtub approach is applied for the Niger delta in order to estimate areas most likely to be inundated by high water levels under SLR conditions.

6.1.1 Date preparation: GIS data processing

GIS data preparation involved making subsets of satellite imagery and SRTM DEM using the area of interest tool box in ERDAS Imagine. NigeriaSat1 images were mosaicked and clipped to extend from the Niger delta coast to 45 km inland. Data enhancement was applied to the LandSat satellite imagery used for land-cover classification; it is de-stripped and histogram equalized to enhance the dominant features for easy classification. Image processing steps used signature editor in ERDAS IMAGINE to record different classes of land-cover identified and the record was used to classify the LandSat imagery. GIS processes used to prepare data for the SLR inundation mapping in ArcGIS 10.1 include: data editing, overlay operations, vector clipping, use of raster calculator for DEM area calculation, Raster to ASCII generation, ASCII to RASTER generation, DEM resampling, use of math and map algebra for inundation calculations, and output map production.

Datasets include: Socio economic datasets on population, gas flaring and oil spills; NigeriaSat1 and NigeriaSatX Satellite imageries; SRTM DEM data, and Google Earth imagery.

6.1.2 Analysis of topography and slope

SRTM DEM data was used to determine the topography and the slope for the Niger delta using ERDAS Imagine 9.1. The SRTM DEM used has a horizontal resolution of 90m and a vertical accuracy of +- 10m, this limitation notwithstanding, the SRTM has proven to be very useful in coastal vulnerability studies as the accuracy improves below 2m for low lying areas (Gorokhovich & Voustianiouk, 2006; Yan, et al., 2015; Jarihani, et al., 2015).

The SRTM DEM was used to generate raster contours for topographic analysis of the Niger delta inland areas. We undertake slope and topography analysis to assess the natural vulnerability to flooding and inundation. Borehole, groundwater consumption, and aquifer data are analysed to determine vulnerability to intrusion of sea salts. Finally, the population distribution data is used to map the potential number of people exposed to these hazards.

6.1.3 Mapping SLR inundation

Flood inundation is mapped here in a GIS environment using water level and topographic data. Following the methodology used by NOAA coastal service centre, U.S.A, the SLR inundated areas were mapped using the modified bathtub approach that accounts for local tidal variability (NOAA, 2012). Bathtub approach says that areas lying under a certain height can get filled with water like a bathtub. In applying the methodology therefore, the study area is flooded with a predetermined water elevation from which the topography of the area is subtracted. In the modified bathtub approach, the water elevation is not uniform as tidal variability of coastal water levels due to tides is taken into account; therefore the water elevation layer consists of a range of values. The modified bathtub approach thus produces results showing areas that are vulnerable given a particular height of water; all dry areas are considered safe, while wet areas are considered vulnerable to inundation. A detailed description of the method is available online (NOAA, 2012).

The data needed to apply the bathtub approach include: SRTM DEM, a tidal surface and SLR values. To generate the tidal surface, the longitude, latitude and water levels of 20 tide gauges located in the Niger delta were used. The data for the tide gauges are given as mareograph generated tide tables (Tides4Fishing, 2016), and the tide Mean Higher High Water (MHHW) values are used to interpolate the raster tidal surface extending inland using ArcGIS (figure 6.1). To inundate the area, Niger delta subsidence values of 7 mm/year and 25 mm/year are added to SLR values of 0.019 m and 0.035 m for the years 2030 and 2050 respectively obtained from IPCC projections (IPCC, 2013, Brown et al., 2011). Thus, with subsidence = 7 mm, SLR for 2030 = 0.14 m (7 mm x period 2013-2030 = 17 years), SLR for 2050 = 0.29 m (7 mm x period 2013-2050 = 37 years); with subsidence = 25 mm, SLR 2030 = 0.44 m, and 2050 = 0.96 m.

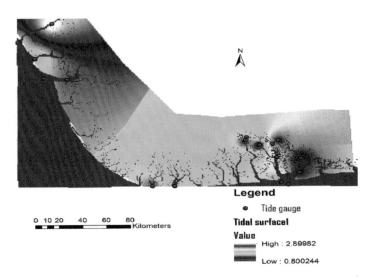

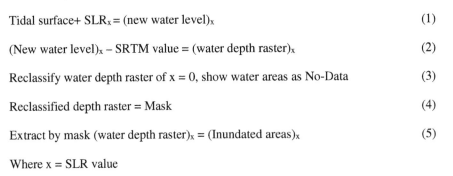

Figure 6.1 Tidal surface for the Niger delta, showing range of tide gauge readings on February 12 2013

After removing all hydrologically[8] unconnected areas, the map of the inundated areas was produced by following the various steps of the bathtub approach. The tidal surface was generated (2013 data) and used as the current water level condition in the Niger delta. The SLR values were added to the interpolated tidal surface raster using spatial analyst extension in ArcGIS to give the new water level. In summary, the methodology is as follows:

Tidal surface+ SLR_x = (new water level)$_x$ (1)

(New water level)$_x$ – SRTM value = (water depth raster)$_x$ (2)

Reclassify water depth raster of $x = 0$, show water areas as No-Data (3)

Reclassified depth raster = Mask (4)

Extract by mask (water depth raster)$_x$ = (Inundated areas)$_x$ (5)

Where x = SLR value

---

[8] The Niger delta coast has low lying locations (e.g. swarms) that are hydrologically unconnected with the ocean, but are naturally wet/inundated. Such areas need to be removed in order to show SLR inundation. Region group and extraction tools in ArcGIS are used to determine those areas connected to the ocean and to remove the unconnected one.

6.2   Results

6.2.1   Local response strategies in the Niger delta

The people of the Niger delta have devised ways of coping with the different disasters that they have been exposed to. The methods used by the people to adapt to these hazards have enabled them to live as best as they can within their environment. For example, communities around the gas flaring sites use gloss paint on their roofs to delay rusting and corrosion (NDRMP, Niger Delta Regional Master Plan: Chapter 1. Niger Delta Region, Land and People, 2004b).

6.2.1.1   Flooding and Inundation

Rising sea levels have directly reached many communities located close to former shorelines. Consequently, many coastal fishing communities have had to raise the floor of their houses and suspend them on stilts to keep above the water (NEST, Reports of Research Projects on Impacts and Adaptation, 2011). In areas affected by frequent river flooding small pedestrian bridges made from wood or sand bags are constructed through community effort (Uyigue & Agwo, 2007). These bridges however have very short lifespans because wood degrades with time, and flood waters erode the sand bags. Where the flood waters are too high and the flood duration is long, residents relocate to safer areas and only return after the floods; however in extreme cases when the houses are damaged and the land inundated, the residents completely abandon their homes (Uyigue & Agwo, 2007). In upper Niger delta communities (around Anambra state) all households are required to have flood receptor pits to collect river flood waters in order to prevent excessive runoff into neighbouring homes and on roads. Similarly, in some coastal communities' embankments, canals and channels are constructed through local efforts to control flood waters and rising sea levels. Farming practices have also been adapted to cope with flooding; farmers now plant on mounds of soil in order to raise crops above flood waters. Other farming methods used include: ridging and terracing of farmlands to form barriers to flood waters, and use of sand filling (NEST, Reports of Research Projects on Impacts and Adaptation, 2011).

6.2.1.2   Erosion

Erosion has been ravaging the Niger delta due to natural causes like river flow and ocean surge, and also by construction of bridges, canals and other coastal structures which altered the natural course of the rivers (NDRMP, 2004a). Depending on the location and local condition different types of erosion exist in the Niger delta, including: gully, bank, ravine, sheet, and coastal erosion. Consequently, oil wells and mangroves (which serve as natural flood protection measures) have been to erosion on the delta (Uyigue, 2009). Under SLR conditions areas that are undergoing erosion will have high vulnerability while areas of accreting sediment will have lower vulnerability (Musa, Popescu, & Mynett, 2014a). To protect the land, the local people use sandbags to fill up eroded areas around their homes; communities located on the coast adapt to beach erosion by constructing walkways and embankments to confine the sea water off the shore. In upper Niger delta communities (around Anambra state) people plant Indian bamboo, Cashew and Banana trees as erosion control measures; other measures include use of gullies as refuse dumps, and planting of grasses. Furthermore, to enable compliance by all residents, community leaders create local legislation against sand mining, tree felling, illegal building, and other practices that expose the land to further erosion (NEST, 2011).

6.2.1.3 Salt water intrusion and contamination of groundwater

For many parts of the Niger delta, groundwater provides drinking water via hand dug wells (NDRMP, 2004a; Shell E. , Environmental impact assessment for the utorogu NAG, 2004). The use of hand dug wells show that the water table is close to the surface; studies in various parts of the Niger delta show an average depth to water level of approximately 10m below the surface. This closeness to the surface exposes the groundwater sources to contamination from the sewage and other contaminants (NDRMP, 2004a; Shell E. , 2004). Although the water in some of these areas is thus contaminated, the people still depend on it for drinking and domestic use (Shell E. , 2004; Shell E. , 2006). This makes their resilience to contamination of the groundwater by intrusion of sea salts to be low.

6.2.2 Inundation mapping

The calculated inundation areas were overlaid on a NigeriaSat1 image of the study area with a total surface area of 24,099 km^2, and the percentage of inundated area was calculated. From the current tidal water levels obtained, a map of the current inundated (February 2013) areas was produced with SLR = 0. The map was compared to classified Landsat image of the Niger delta of same period to check for the present inundated and low lying areas. The classified Landsat image (figure 6.2-right) shows low lying wetland areas regularly inundated at high tide and during the rainy season. Such areas can be dry at low tide and can be used as farmlands; in the coast such areas are usually vegetated with mangroves and heavily forested.

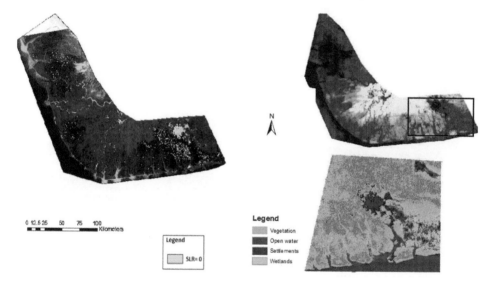

Figure 6.2 February 2013 coastal conditions (left) Inundated areas with SLR=0, (right) Landsat image of the Niger delta, the area within the box is classified as shown.

With 7mm/year of subsidence, the results for 2030 show that a rise in sea levels of 0.14m will cause an inundation extent of 1119.3km^2 which is 4.6% of the total surface area; while subsidence of 25mm/year, giving SLR of 0.44m 2030, causes an inundation extent of 1254 km^2 which is 5.2% of the surface area.

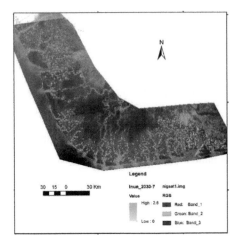
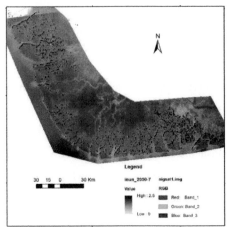
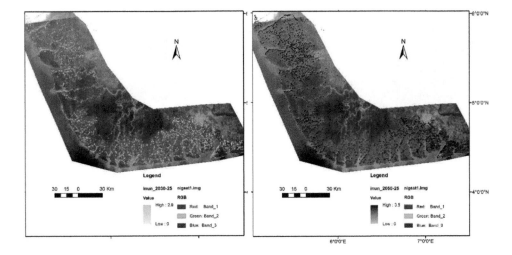

Figure 6.3. (Above) Inundated areas by 2030 with SLR=0.14m and 0.44m; (below) inundated areas by 2050 with SLR=0.29m and 0.96m.

Results for 2050 show that a rise of 0.29m (subsidence=7mm/year) will cause an inundation extent of 1175.9 km² which is 4.9% of the total surface area. With SLR of 0.96m (subsidence=25mm/year) by the year 2050, the inundation extent is 1633 km² which is 6.8% of the surface area.

From the inundation maps (figure 6.3) a rise in sea levels of 0.14m already inundates more areas in the Niger delta, and this area further increases with SLR of 0.96m. Inundated water depths also increase with higher SLR values, thus areas already inundated at lower SLR are likely to have deeper water depths with increase in sea levels.

6.2.3 *Physical characteristics of the Niger delta*

6.2.3.1 *Topography*

The elevation of an area determines the lowest level of water that can flood it. Thus it controls which part will be impacted by effects of rising sea levels like storm surge and coastal flooding. Low lying areas have a higher likelihood than higher areas of getting flooded during a storm surge.

Using SRTM DEM, contour lines were generated for the Niger delta at 1m contour interval in ERDAS Imagine 9.1. The Niger Delta from coastline to 45Km inland was classified for elevations from 0 to 10m (all other areas were excluded before classification). Figure 6.4 shows the elevation range of areas from 0 to 10m above sea level. The elevation range shows that there are inland areas with very low elevation which makes them vulnerable to the effects of SLR like flooding and inundation (Gornitz, Couch, & Hartig, 2001; Musa, Popescu, & Mynett, 2014a).

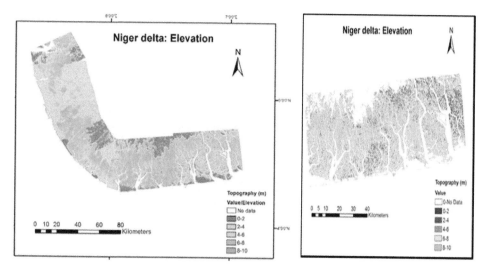

Figure 6.4 Topography of the Niger delta from coastline to 45Km inland, showing areas with elevations from 0-10m. Topography classes show land areas lying almost at sea level; thus flooding from high tides and storm surge is easy

6.2.3.2 Slope

The slope of an area controls the flooding extent so that areas with steeper slopes offer more resistance during a flooding event than those with gentle slopes. The Coastal areas with gentle slopes are therefore more vulnerable to rise in sea levels. To determine the slope of the Niger delta, SRTM DEM was used to generate slopes in percentage rise from the coastline up to at least 45km inland (figure 6.5). The resulting slope map shows that there are many areas with 0-1% slope located inland making them very vulnerable to the effects of SLR like flooding and inundation from storm surges. The slope data generated was verified using the slope map of Nigeria accessed from NASRDA archives.

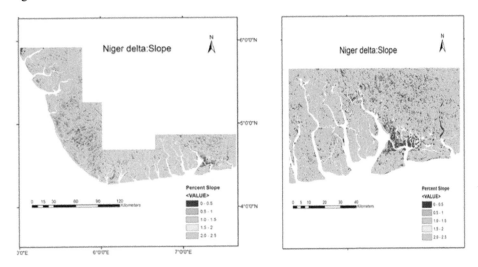

Figure 6.5. Slope range of the Niger delta; shortest distance from shore is 45km inland. Slope values show the Niger delta land area are almost flat, thus increase in inundation extent from flood waters will be easy.

6.2.3.3 Borehole and Aquifer conditions

Due to the likely effect of inland intrusion of sea water when the sea level rises, the type of aquifer in a given area determines how vulnerable its ground water will be to salt water intrusion. The Niger delta has confined, unconfined and semi-confined aquifers with different properties and suitability for domestic use (NDRMP, 2004d).

Confined aquifers are overlain by materials with poor permeability which limit inflow and outflow, whereas unconfined aquifers are overlain by highly permeable materials with open pore spaces which allow direct interaction with the surface and makes them easily accessible to contamination; semi-confined aquifers which have a layer of relatively low permeability above them but still allow vertical flow of water (DPIPWE, 2012). The Niger delta has eight different geological formations: the Crystalline Basement Complex, Cross River Group, Ajali Sandstone Formation, Nsukka Formation, Bende-Ameki Formation, Benin Formation, Deltaic Plains Formation and the Alluvial Deposits.

The areas closest to the Niger delta coastline are found within the Alluvium, Deltaic Plains and Benin Formations which contain water with high salinity.

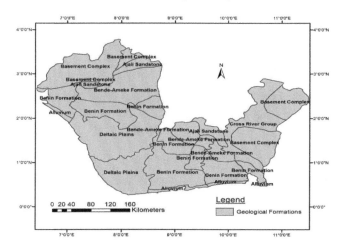

Figure 6.6. Geological formations in the Niger delta. Aquifers located on Deltaic plains, Benin formation and Alluvium are salty

The people living in those areas therefore use surface water and rain water as the main sources of drinking water as the groundwater is not drinkable (NDRMP, 2004a). For areas above the immediate coast, there are settlements that depend on borehole water as the primary source of drinking water; such areas will be vulnerable to contamination as a result of intrusion of sea salts.

6.2.3.4 Population distribution

The population of the Niger delta increased by an average growth rate of 3.1% between 1991 and 2006 (NPC, 2010). With such increase in population, more people are exposed to hazards in the delta. Population increases the value of risk an area is exposed to because human settlements come with infrastructure, farming, and other economic activities which can be affected by hazards.

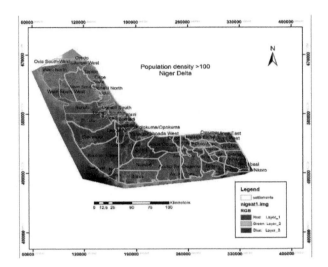

Figure 6.7. Population density of the Niger delta. Highlighted blue boundaries show local government areas with high population density.

Settlement data for the Niger delta comprising point data of Nigerian towns and Google earth satellite imagery were used to determine the presence of settlements located within a distance of 0-1500m of the coastline. Results showed 320 of such settlements. Due to their closeness to shore, these settlements are very prone to flooding, inundation and storm surges.

Using the population data of the Niger delta per local government (LG) area, the population density of the Niger delta was calculated as Pop-density = Population/LG area. (Although this method assumes equal distribution of people within a land area which is not usually the case, in the absence of population data per settlement it is used here).

Figure 6.7 shows the densely populated parts of the Niger delta with greater than 100 people per square kilometre. These areas are mostly located along the coast and the river banks (NDRMP, 2004b). Since fishing is a very important economic activity many towns and villages are located very close to the coastal waters.

6.3 Discussion of results as they relate to local adaptation practices in the Niger delta

The analysed slope and topography of the Niger delta (figures 6.4 and 6.5) show that it is vulnerable to further flooding, inundation, and consequently erosion as a result of sea level rise. The results of GIS inundation maps (figure 6.3) also show increase in inundation extent and depths with SLR; this implies that more residents will be affected by flooding and inundation. Furthermore, the aquifers located in the Niger delta coast are salty and therefore undrinkable; with high population density in the coastal areas, there is high vulnerability to SLR intrusion through inundation of surface water. Hence, the Niger delta needs implementation of sustainable adaptation and mitigation plans that will protect the land and people. These plans can include the local methods already familiar in the Niger delta, however some of the practices might not be sustainable. The advantages and disadvantages of the present practices are thus discussed below.

- Utilization of flooding 'pits' to control runoff as practiced in the upper Niger delta has been helpful in creating small water reservoirs that are later used for other purposes. This practice however, has two disadvantages: first, the reservoirs can serve as mosquito breeding grounds if not well built and protected from dirt; secondly, impounding flood waters impedes the flow of sediments to the downstream delta area thus decreasing soil replenishment which can affect the delta. Decrease in sediment replenishment makes deltas to subside, therefore increasing their vulnerability to effects of SLR (IPCC, 2007b). With the inundation maps (figure 6.3) showing further inundation of inland areas with SLR, the use of flood pits should not be an option for areas within 45 km of the shore. This is because the area will likely be inundated and any flooding will be predominantly coastal; moreover, the concept of storing water using flood pits will not be needed for salty sea water. However, since use of floodpits is a simple and effective method for flood runoff retardation (and improves the coping capacity of the people), areas located upstream where flooding is predominantly fluvial can use floodpits as an adaptation method. The quality of the pits should however be controlled to avoid contamination and impeding of sediment transportation downstream.

- Under SLR conditions areas that are undergoing erosion will have high vulnerability while areas of accreting sediment will have lower vulnerability (Kumar & Kunte, 2012). As the common practice against erosion in the upper Niger delta is to dump refuse inside eroded gullies in order to fill them, part of such refuse will be washed and deposited downstream during heavy rainfall and flooding. Since the underground aquifers in the coastal areas of the Niger delta are salty and the people depend on surface water as the primary source of drinking water (NDRMP, 2004b) (NEST, Reports of

Research Projects on Impacts and Adaptation, 2011), this practice should not be encouraged as downstream areas can be contaminated and surface water quality affected.

- Other practices like planting bamboo trees, bananas and grasses to reduce erosion have several advantages including: providing a good land cover that protects and binds the soil, providing extra source of healthy food and income for the people, and improving the biodiversity of the area. These practices should be encouraged and introduced in areas around the upper Niger delta where the water is fresh. However, in areas found within the Alluvium, Deltaic Plains and Benin Formations (figure 6.6), bamboo trees cannot be used as an adaptation strategy because bamboo plants do not grow well (and even die) in salt water (All science fair projects, 2015).

- Figure (6.3) shows that SLR will bring sea water further inland, much closer to farms and settlements. The practice of raising building floors above flood water levels being practiced in the immediate coastal communities, can therefore be a useful solution for other vulnerable areas further inland. This method can be adapted into new building plans (for example) to avoid future inundation.

- As practiced now in areas in the Niger delta, sandbags can also be used as embankments to keep water out of built up areas if enough sandbags can be provided for the local people; however this practice encourages sand mining which accelerates erosion and should therefore must be regulated.

- For seasonally flooded areas, use of 'temporary' bridges can be made more effective if better materials than wood are used to construct the bridges (e.g. metal).

6.4 Conclusion

Analysis of Niger delta slope and topography show that with sea level rise, the Niger delta is vulnerable to further flooding, inundation, and erosion. Using SLR values of 19mm by 2030 and 35mm by 2050 (in addition to estimated subsidence values), this chapter shows that Niger delta can lose 4.6-5.2% (1119.3 -1254km^2) of its land area to inundation by 2030, and 4.9-6.8% (1175.9 -1633km^2) by 2050. Coastal waters moving inland means the future inundated areas are also vulnerable to erosion and storm surges by SLR. The results for inundation modelling also indicate that without subsidence the inundation effect of SLR on the Niger delta will be minimal since the eustatic values are just 19mm and 35mm by 2030 and 2050 respectively. However, subsidence has made the Niger delta very vulnerable to inundation by making the RSLR values very high. Since eustatic SLR is predicted to reach 1m by 2100, and oil and gas exploration in the Niger delta continues, there is higher risk of further inundation and loss of land. Thus, the Niger delta needs adaptation and mitigation measures against SLR.

Present local practices used within the Niger delta have helped the locals to cope with challenges posed flooding inundation, erosion, and groundwater contamination; however with a rise in sea levels these hazards will increase in frequency and intensity (IPCC, 2013). One important effect for the Niger delta is that high sea levels will impede the draining of river waters, so that more upstream areas in the Niger delta will be inundated from flooded rivers, and water depths will increase in estuaries (Musa, Popescu, & Mynett, 2014b). The resilience shown by the local people can thus be further strentgthened by adopting some of the sustainable local practices used by them as adaptation strategies to effects of SLR. Sustainable practices are those that add to the resilience of the people and do not negatively affect the environment, biodiversity and community life of the people. Sustainable local practices in the Niger delta include: planting of Bamboo trees for erosion control, use of sandbags as bridges and dykes (flood control), use of flood receptor pits as temporary flood water reservoirs, and community legislation against sand mining and indiscriminate tree felling. Other good practices undertaken by the people such as sinking of boreholes, construction of canals and embankments, are however not sustainable as they are generally below standard and undertaken without environmental assessment studies (NEST, 2011). There is therefore a need for the government to undertake such projects and ensure they are done properly.

Finally, the Niger delta is rated low for availability of effective emergency services, disaster communication and shelters, this reduces the resilience of the people (Musa, et al., 2014). These factors are directly provided by the government at the state and local levels, therefore more needs to be done in training and enabling of local volunteers especially women who constitute half of the population and are most vulnerable (NBS, 2011).

7

Mitigation and adaptation

to sea level rise [9]

[9] This chapter is an edited version of the submitted paper:

Musa, Z. N., Popescu, I. & Mynett, A., 2018. Developing adaptation plans for Nigerian coast based on around the world experiences. Accepted for presentation at River Flow 2018 conference, Lyon France.

The biggest problems of planning for the various consequences of climate change include: the limitation in data availability about the climatic conditions of many areas, the interactions within the system, and the many phenomena that affect the behaviour of the system in response to the changing conditions. Another problem is that many coastal areas are highly populated and their morphology is modified by anthropogenic factors which increase the uncertainty of climate change predictions. These problems make it difficult for particular solutions to be adopted as the best options for adaptation/mitigation for a coastal area (Laukkonen, et al., 2009). Accelerated SLR is the most important climate change impact for coastal areas, consequently, one of the three working groups first established by the IPCC is the Response Strategies Working Group (RSWG), which was mandated to examine options for limiting climate change and adapting to the changes (Dronkers, et al., 1990). Thus the RSWG was mandated to propose adaptation and mitigation options to manage SLR.

Adaptation and mitigation apply to two different spatial scales. Structural mitigation measures against SLR are quick short term solutions that are effective against hazards and mostly applicable at a national level. Adaptation measures are long term solutions that are more applicable at the local level. Mitigation options are capital intensive measures like: shoreline armouring with hard structures like groins, seawalls, jetties, breakwaters; measures which generally intercept wave energy/reduce movement of sand and therefore also undergo erosion at their base and need to be rebuilt (Gornitz, Couch, & Hartig, 2001). Flood adaptation strategies strengthen communities in building their own coping capacities that shape the local climate change policy (Uy, Takeuchia , & Shawa, 2011). For example, in the case of a vulnerable urban area, adaptation strategies can be applied in spatial planning of new urban areas, or to enhance flood protection. In the first case the amount of new urban land in flood-prone areas is limited by enforcing building restrictions, while in the second case flood defences are enhanced to withstand higher and more frequent flooding (Muis, Güneralp, Jongman, C.J.H. Aerts, & Ward, 2015). SLR adaptation measures for coastal areas include: floating agriculture, buildings suspended above flood waters, roads exchanged for water ways, resettlement, changes in land use planning, planning of flood resilient settlements etc. (Rijsberman, 1996).

Over the years several countries have committed to non-structural climate change mitigation measures like CO_2 reduction, green roofs and afforestation. Although these mitigation strategies reduce the severity of a hazard, the ocean's slow response to climate change mitigation measures and the high cost of SLR mitigation, makes adaptation to sea level rise preferable (TOL, 2007); moreover, the lack of infrastructure to cope with the effects of climate change make adaptation the best option for developing countries. Adaptation is the ability of a system to adjust to new unfavourable conditions in such a way that it minimizes the negative effects of potential damages. A systems adaptive capacity enables it change is characteristics or behaviour so that it can cope with anticipated or existing hazards. When a system adapts it directly reduces the people's vulnerability to the hazard (Brooks N. , 2003), and strengthens its existing coping strategies (Uyigue & Agwo, 2007). Therefore for sustainable solutions to climate change issues, a combination of adaptation and mitigation measures can provide better outcomes (Brooks N. , 2003; Laukkonen, et al., 2009).

This chapter considers possible mitigation and adaptation to SLR measures. The analysis is divided into two main parts; the first assesses the types of adaptation plans that can be made for deltaic areas, it proposes adaptation scenarios, possible options they entail, and the implications of each. In the second part, some of the proposed adaptation options are applied to the Niger delta and their effects simulated and discussed.

7.1 Mitigation and Adaptation options for deltas

The IPCC Response Strategies Working Group (RSWG) recommended three strategies for response to SLR: 'Retreat', 'Accommodate' and 'Protect' (Dronkers, et al., 1990). Under 'Retreat', no effort is made to protect the coastal area, it is abandoned and left for natural processes to continue. 'Accommodate' option allows the people to continue living within the threatened area without making efforts to stop natural processes (e.g. flooding) from taking place; thus the people adjust their lifestyles without disturbing the ecosystem. The option to 'Protect' entails making efforts to control the natural processes that come with rising sea levels (flooding, inundation, erosion etc.) through construction of hard structures like dykes, and applications of soft solutions like planting of trees/vegetation to protect the land from the sea waters.

To implement each strategy, different conditions are required. A 'Retreat' strategy for example, will depend on availability of alternative lands; an 'Accommodate' option will require capacity building (in e.g. flood management) and availability of flood insurance, a 'Protect' option requires availability of funds and dedication to long term plans. However, none of the three options can be said to be the 'ideal solution' that is universally applicable; this is because most times, applied response strategies are mixtures of the three options (retreat, accommodate and protect).The response strategy implemented in an area depends on the value of the coastal area, the political and local considerations, local geo-morphology, ecology, and the cost of the adaptation measures to be applied (Tol, 2007). Depending on the decisions taken by the authorities and stakeholders, the adaptation option used has an implication on available methods that can be used and cost of the adaptation plan. Furthermore, each option demands different levels of commitment from the government and stakeholders.

Response Strategies

Considering these response options ('Retreat', 'Accommodate', and 'Protect') possible adaptation strategies for deltaic areas can include the following:

Under retreat:

- Plan resettlement of vulnerable communities to safer (less vulnerable) areas
- Stop further infrastructural development projects in vulnerable areas

Under Accommodate:

- Allow flooding/inundation/ erosion of low lying areas
- Resilient settlements planning (all new plans for infrastructural developments must include resilience e.g. building/ city plans, farming methods)

Under Protect:
- Apply hard structural measures (e.g. dykes, gates, groins in all vulnerable areas)
- Apply soft structural measures (e.g. planting of vegetation in all vulnerable areas)

Implementation of each adaptation strategy affects the cost of adaptation. For example, a plan to 'allow flooding/inundation of low-lying areas' might include costing of loses that will be incurred by farmers using such lands. A plan for resettlement might relocate residents to a new area or just pay compensation to those migrating. The most expensive measures/plans however are: Resettlement of residents, planning of resilient cities, and application of structural measures.

Planning Extent

The extent of an adaptation plan can be all encompassing or limited to areas of special interest (figure 7.1). The adaptation extent has an effect on the types of measures applied, the possible beneficiaries/stakeholders, as well as the cost of implementing the adaptation plan. For example, limiting an adaptation plan to economically viable areas/ areas of special interest (e.g. wetlands), gives such areas priority irrespective of other sector vulnerabilities; besides such a plan will most likely be a 'protect' strategy.

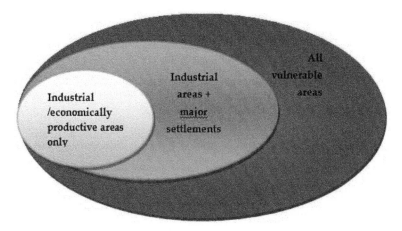

Figure 7.1. Possible adaptation coverage areas for a delta. The area of coverage limits the adaptation cost, type and beneficiaries.

On the other hand a plan made to cover both economically viable areas and settlements will include options under Retreat, Accommodate and Protect (figure 7.2). An adaptation plan covering all vulnerable areas will include processes that will intervene on all areas; consequently all natural processes will be affected.

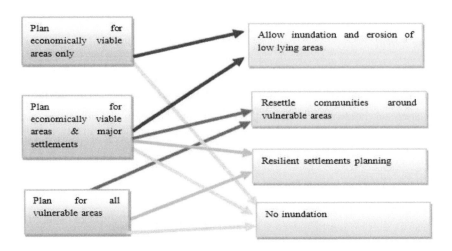

Figure 7.2. Possible adaptation options available under different adaptation plans

7.2 Options for the Niger delta

The risk-based Netherlands Delta program approach allows for a wide variety of flood risk measures to be used; by considering flood consequence reduction in addition to flood probability reduction, and allowing different standards to be set per individual dyke ring (Nillesen & Kok, 2015). Nillesen & Kok, (2015) applied the risk-based delta approach on the Netherlands on the Alblasserwaard dyke ring to determine the best flood risk reduction strategy. Their results showed that an integrated approach that combines different interventions provided the best solution against flooding; it recommends use of by-pass channels for river level reduction, broadening of a highway along a canal, and less dyke reinforcement (thus maintaining the spatial desirabilty of the land). Following Nillesen & Kok (2015), different possibilities of reducing the adverse effects of SLR on the human population are tested for the Niger delta using the data obtained from the various methods used throughout this study.

The data are used as inputs and / or decision criteria, as follows:

- In chapter 4, models of river and coastal flooding were run to simulate the effects of SLR, and the results showed possible impacts on the river dominated and tide dominated coasts respectively. The data generated are used as criteria for impact reduction (i.e. water level, water depth, flood/inundation location, flood reach, and flood/inundation duration).Thus coastal interventions can be added to the existing models and the results compared with those obtained in chapter 4.

- The risk-based approach combines flood probability modelling with consequence assessment and sets safety of human life as the standard. In chapter 5 the calculated $CV_{slr}I$ for the Niger delta coastline prioritised human interventions and properties in ranking vulnerability, the results of the $CV_{slr}I$ are used in two ways: (1) to locate areas needing intervention, (2) to determine the consequence criteria (resilience index, susceptibility index).

- In chapter 6 the resilience practices being used in various parts of the Niger delta were assessed for their sustainability if proposed as future adaptation measures. The applicability of those measures deemed sustainable are tested here by simulating their impact on the hydrodynamic model results.

7.2.1 Scenarios for mitigation/adaptation

Based on SLR adaptation strategies of 'Retreat', 'Accommodate', and 'Protect', we consider four possible options for the Niger delta: 'Do nothing', 'Resettlement', 'Plan resilient settlements', and 'Allow no inundation'. These options entail taking certain decisions which implement actions that reduce the risk of SLR for an area; the steps to follow are presented in the following scenarios:

Scenario 1: Do nothing

- Continue with existing practices (business as usual)

Scenario 2: Resettlement

- Locate empty (available) suitable land
- Remove small settlements (low population) to safer areas where possible
- Calculate cost implication of relocation, and budget for compensation

Scenario 3: Plan resilient settlements

- Put dykes around vulnerable settlements
- Dredge channels
- Allow flooding of low lying areas (farmlands and floodplains)
- Change farm practices?
- Introduce by-pass channels and where there's river flooding add reservoirs/ detention basins (with gates/sluices that can slowly drain the water after the rainy season)
- Planting of Mangrove trees on low-lying area
- New building plans on raised ground levels

Scenario 4: Allow no inundation

- Dyke all vulnerable parts of the coast based on physical properties
- Construct sea defence structures (groins, sea walls, etc.)
- Dune nourishment
- Dredge channels
- Bypass channel/ reservoir (for river floods)

7.2.2 Implementation Criteria

The conditions under which an adaptation strategy is applied to an area are given in table 9. Since the 'Do nothing' scenario maintains the scenario status quo whereby conditions in the Niger delta are left to continue as they are, and respond to natural and anthropogenic effects without an integrated SLR adaptation strategy, no planning criteria is set for it.

Table 9. Criteria used for implementation of different mitigation/adaptation strategies for the Niger delta

No	Scenario	Criteria
1	Do nothing	None
2	Resettlement	- $CV_{slr}I$ = high – very high - GIS inundation map = inundation even at lowest level SLR + subsidence - Hydrodynamic model = high water depths/levels, earlier occurrence of flooding, and extension of flood reach upstream - Settlement size is small/ fishing village - Settlement type is rural (no coastal infrastructure)
3	Resilient settlement	- $CV_{slr}I$ = medium – high - GIS inundation map = inundation inundated by 2030 with level SLR + 25mm/year subsidence - Hydrodynamic model = high water depths, plus increasing upstream/lateral flooding extent - Settlement size is large/ high population density - Settlement type is town
4	No Inundation allowed	- $CV_{slr}I$ = medium - very high - GIS inundation map = inundation even at lowest level SLR + subsidence - Hydrodynamic model = high water depths/ water levels, increasing lateral flooding extent - Settlement type oil and gas infrastructure - Coastal infrastructure present

7.2.3 Coverage area and planning extent

Based on the classification of vulnerability as computed and mapped in the $CV_{slr}I$ (coastal vulnerability to sea level rise index) in chapter 5, the parts of the Niger delta coast most in need of mitigation and adaptation plans are as shown in figure 7.3. As a tide dominated area that is densely populated, and also has oil infrastructure, area C will be affected by SLR (as shown in section 4.2). The adaptation strategies proposed in this chapter are demonstrated on area C.

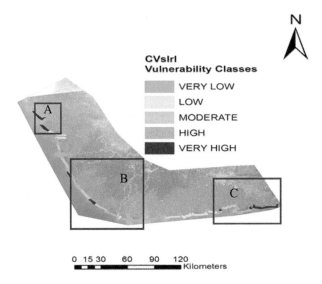

Figure 7.3. Showing boxed areas (A, B, C) that have medium to very-high CVslrI ranking and need SLR adaptation plans.

As stated in section 7.1, the planning extent determines the type of strategy that can be applied, the area that can be covered, and the beneficiaries. Adaptation/mitigation strategies can be applied under three possible planning extents for the Niger delta (figure 7.4).

1. To cover all vulnerable areas (as mapped in the CVI)

2. To cover only areas around oil and gas facilities/ reserves

3. To cover areas around oil and gas facilities/reserves + vulnerable major settlements

In this chapter the adaptation strategies will be applied to cover all vulnerable areas within the modelling domain.

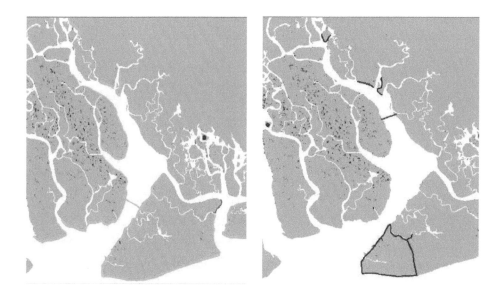

Figure 7.4. Adaptation planning extents for the Niger delta. (Left) Protect all vulnerable areas. (Right) Protect Oil/gas infrastructure and major settlements only.

7.2.4   Effects of measures on the study area

Just like the case of coastal Louisiana and the Netherlands, a long term protection against possible effects of SLR should include hard structures like dykes as well non-structural measures like beach nourishment and maintenance of practices that encourage sediment replenishment for the delta.

According to the implementation criteria in table 9, the adaptation strategies that can be applied to areas inland of coastline 'C' (with medium to high vulnerability) are 'Resilient settlement planning', and 'No inundation'. Resettlement is not considered a viable option for area C because it has large densely populated towns, and presence of oil and gas infrastructure. The methodology applied to implement the adaptation strategies are summarized in table 10.

Table 10. Implementation of the adaptation strategies based on available data

No	Inputs	Decision/ Plan	Approach	Output
1	Modelling data, DEM/TIN	Do nothing. Business as usual.	-Create $^+$LU/LC map, Manning's table of rough roughness values/LU, DTM from water depths. -Calculate $CV_{slr}I$ -Setup and run model 1 - Calculate cost of loss	-$CV_{slr}I$ -Model 1* results -Cost -Mitigation/ Adaptation option
2	DEM/TIN, flood map, $^+$LU/LC map, modelling data. Model 1	Resettlement.	-Check LU/LC map for vacant land -Check DEM heights of vacant land -Move settlement to new site -Setup and run model 2 -Calculate $CV_{slr}I$ -Calculate cost of resettlement	-New $CV_{slr}I$ -Model 2** results -Cost
3	DEM/TIN, flood map, LU/LC map, modelling data. Model 1	Resilient settlement planning.	- Put Dykes/ Canals around settlements - River dredging (Rivers/ estuaries) -Add river bypass connected to reservoirs/ retention basins (River dominated coasts) -Allow inundation of low-lying -Setup and Run model 3	-Model 3*** -Cost -Mitigation/ Adaptation option
4	DEM/TIN, flood map, LU/LC map, modelling data. Model 1 $CV_{slr}I$	No inundation	-Dyke entire coastline around vulnerable areas - Dune sand nourishment / sea wall construction -Dredge channel (River dominated coasts) -Add a surge barrier in the estuary -Setup and run model 4	-$CV_{slr}I$ -Cost -Mitigation/ Adaptation option

+Land-use/ land-cover map. *Model 1: model results from chapter 4. **Model 2: output of modified model with settlement moved. ***Model 3: output of model modified by adding coastal interventions (e.g. dykes, surge barriers).

Scenario: Plan resilient settlements

Under this scenario, the resilience of existing occupied/utilized lands is improved by applying diverse types of measures. These measures include those that will protect the land from the flooding/inundation effects of SLR, those that will increase the carrying capacity of the rivers/creeks, and those that reduce the properties exposed to impacts of SLR. Under this scenario therefore the following conditions are simulated and their effects on flooding, inundation and erosion indicate how effective they will be against SLR.

Use of dykes/embankments

These involve use of hard structures along the vulnerable areas and shoreline (revetment/seawall) to levels where they can stop high water levels from flooding/inundating an area. Dykes are usually constructed to fit some design flood level calculated based on historical data, in the Netherlands for example all Sea defence dykes are designed to withstand a 1 in 10,000 year storm water level (7.65m above NAP). The most significant storm surge recorded in Bonny river raised water levels 50cm above Spring tide levels (NDRMP, 2004a); with SLR plus land subsidence values reaching 0.96m by 2050 for the Niger delta (section 4.1.1), storm surge water levels can therefore rise to 1.5m above normal. The D-flow coastal flood model setup in section 4.2 was run with 1.5m SLR and the model simulation results show that water levels in Bonny River can rise up to 5m with 1.5m SLR; consequently and coastal protection should be able to withstand at least 5m of water level.

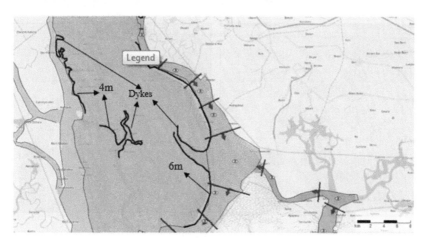

Figure 7.5. Dykes around vulnerable areas along Bonny River and inner flow channels Niger delta.

To simulate this, dykes are added to the D-flow coastal flood model; the dykes are put around areas that showed significant flooding in the modelling results (figure 7.5). Different dyke crest levels were used to simulate the protection level, and the optimal dyke crest heights that minimised flooding (up to 1.5m SLR) were 6m for areas directly connected to the sea and 4m for inland areas (figure 7.6).

The simulation results (figure 7.6-bottom), show that application of dykes have stopped flooding of areas close by. This is important as the modelled area has several settlements that are at risk of flooding, adding the dykes will help lower this risk. As discussed in chapter 6, some local communities in the Niger delta use sandbags for flood and erosion control, this knowledge will be very useful as sandbags can also be used to construct cheap embankments by individuals where large dykes are unavailable.

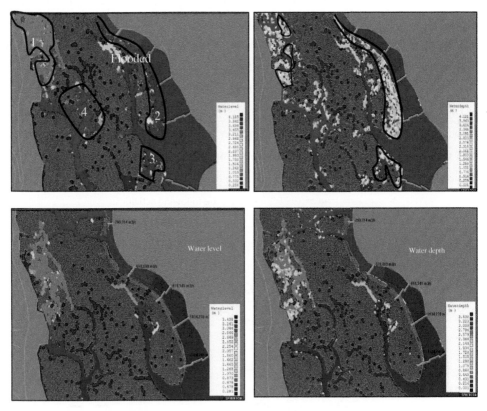

Figure 7.6. (Top) water level and water depths on flooded areas (1, 2, 3 and 4) with 1.5m SLR. (Bottom) water level and water depths results after applying the Dykes as shown in figure 7.5.

Bypass channel

Area 1 (figure 7.6 - top) with several vulnerable settlements still has flooded areas even after dykes were applied along the channels (figure 7.6 - bottom). Model simulation shows that the flooding came from Bonny River through one of the channel connections (figure 7.7b). Model simulation results on the channel cross section (figure 7.7b) showed flow discharge values up 90m^3/s. A bypass channel with an incremental discharge capacity of 40 - 100m^3/s was therefore added to redirect the flow into the New-Calabar River.

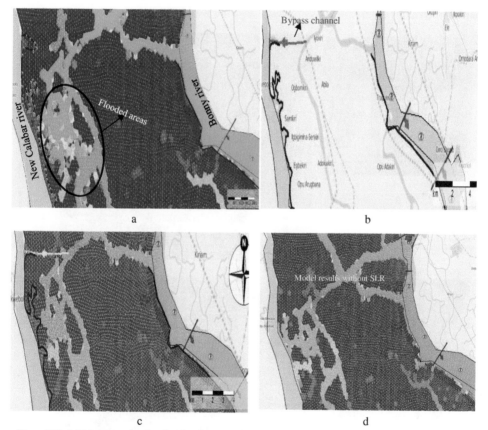

Figure 7.7 . (a) Model results showing flooding on area 1 even after addition of dykes. (b) Map showing the channel through which Bonny river floods area 1 and the position of the added bypass channel. (c) Model result showing how the effect of the bypass channel on flooded area extent has reduced it to levels comparable to model flood extents with no SLR (d).

Model simulation results showed that the bypass channel reduced the area flooded by Bonny River to extents comparable with simulation results without SLR (figure 7.7d). Use of bypass channels relates to the use of flood pits by locals as a flood resilience practice (chapter 6)

Channels dredging

Making the rivers deeper will increase their carrying capacity (although this is more desirable to protect against river flooding) so that higher water levels will not overflow the banks. Similarly, widening the channels where possible will improve carrying capacity. To simulate channel dredging, the model was set up with the main dykes removed, and the estuary and river depths lowered by 2.5m.

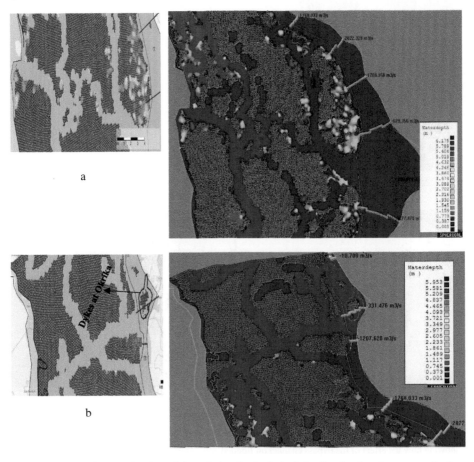

Figure 7.8. Effect of dredging the channels by reducing the depths by 2.5m. Results show lower water depths on flooded land areas, and high water levels in the channels.

The results were checked for channel increase in water depths, and decrease in flooded area. High water depths values for a 1.0m SLR increased from less than 4m (results without dredging) to over 6m in the modelling domain. The results show more water contained within the channels and less flooding/ lower water depths on the land areas. However, the dredging results indicate that where there are dykes (e.g. in Okrika as shown in figure 7.8b), the high channel water levels will overflow the dykes (unless the dykes are made higher than 6m). Dredging might therefore not be such a good choice for SLR mitigation and adaptation as it does not stop flooding of inland areas, and is likely to cause breaching of dykes (where they exist).

Allow flooding of low-lying areas

The most important criteria for coastal area protection in this study is the presence of human settlements and investments, however the modelling results show some uninhabited that are flooded with SLR. Based on the human habitation criteria, such areas will therefore be allowed to flood. The areas can however be identified in future adaption/mitigation plans as places that should not be inhabited in future (figure 7.9). As mentioned in chapter 6, this concept might not be new as some parts of the Niger delta already have local legislation against practices that encourage erosion, people are therefore already aware of the need to help save the environment. However enlightenment campaigns and payment of adequate compensation to owners of such lands will encourage compliance to the rules.

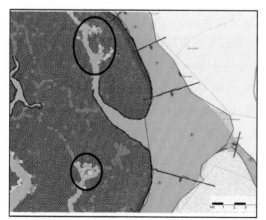 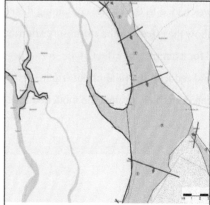

Figure 7.9. Showing uninhabited low-lying areas circled. Locations like these that have no human settlements of infrastructure can be left to get inundated.

Scenario: No inundation allowed

Under this scenario, all areas covered by the adaptation plan will be protected with hard structures to prevent any possibility of flooding. Thus the measures applied under the "Resilient settlements planning' scenario will be applicable with the exception of low-land inundation and wetlands restoration. To further reduce possibilities of flooding or inundation, this scenario will also include use of storm surge barriers, seawalls and revetments, as well as coastline shortening.

Storm surge barrier

Storm surge barriers as implemented in the Netherlands and Louisiana (USA), protect coastal areas from flooding and high waves resulting from extreme weather phenomena like storm surges and hurricanes. Although the Niger delta has only experienced storm surges that raise the water level by less than 1m, with SLR the storm surge levels might be raised to orders of magnitude higher than ever recorded. To simulate this, a storm surge barrier was added into the D-flow model at the location shown in figure 7.4 (left). Although this choice of location leaves out the town of Bonny, surge barriers are built inside the channel towards the inland (e.g. the Maeslantkering is 5km inland); besides Bonny is highly urbanised and it will therefore be difficult to find enough (and available) space for construction.

Although the most suitable barier height has to be tailor made for the Niger delta, a 22 m high storm barrier was used in this model (following the example of the Netherlands). The barrier was added to the model containing dykes as used in previous simulations. The model results show the effects of the storm surge barrier on water levels in Bonny River. At observation point 8 for example, the effect of the storm surge barrier has lowered water levels in the in the channel and reduced the tidal effect (figure 7.10). This result shows that adding a storm surge barrier can effectively protect the modelled area from extreme coastal events.

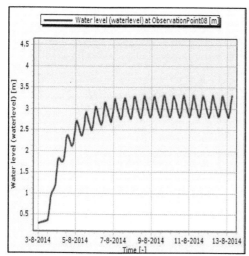
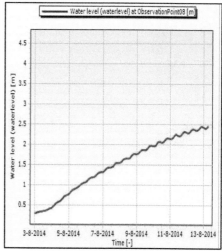

Figure 7.10. (Left) model water levels with 1.5m SLR at an observation point. The effect of a storm surge barrier on the water levels (barrier is present all through model duration).

Coastline shortening

One of the flood protection methods used by the Netherlands is shortening of estuaries and tidal inlets. The modelled area consists of smaller estuaries that also cause flooding of inland areas; these inlets can be shut to reduce the effect of SLR on the inland areas. To simulate this, the model was setup with the inlet shown in figure 7.11 shut using a 10m high dyke. The results show that shutting the inlet has dried up the channel thereby preventing flooding of inland areas. This effect has the advantage of making more dry land available, and reducing the cost of dyke constructions to stop flooding in the inner channels (see figure 7.5 for inner dykes). However, the channel supports a number of settlements (circled area in figure 7.11), and closing it will displace citizens living there. Thus this option can best be applied after consultations with stakeholders and planning of resettlement areas, insurance, and compensation of voluntary migrants.

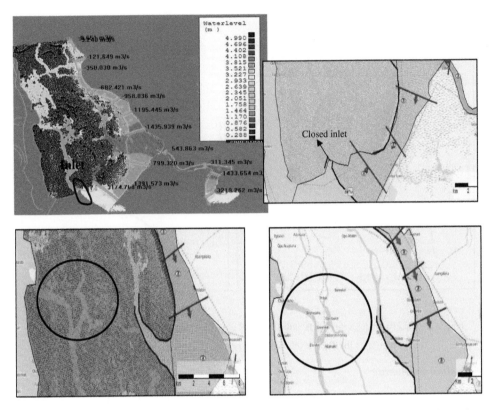

Figure 7.11. Effect of tidal inlet closure on the model results. (Above) The closed inlet. (Below) Model result showing the dried up channel after closure (left), before closure (right).

Seawalls and Revetments

The 'No inundation scenario' also necessitates protecting coastlines by applying seawalls, groins, revetments etc. Since the main parts of the modelling domain is around an estuary, and the flood waters (from our model) are carried via Bonny River, applying seawalls will not be effective against SLR. Therefore this flood defence possibility is not modelled in this chapter. Options like beach nourishment are also not applicable as the geomorphology of the area is deltaic with no beaches (Awosika & Folorunsho, Nigeria, 2012).

7.3 Cost of implementation

Coastal protection is very expensive in monetary terms, however the cost of inaction will be both monetary and non-monetary; threatening loss of lives, properties, history, traditions and cultures. In preliminary efforts to calculate the cost of mitigating climate change effects (including SLR) for the Nigerian coastal region, an estimate of $5b was given for coastal stabilization and resettlement of environmental refugees. Since a comprehensive adaptation cost estimate was yet to be done, the estimate was based on a business as usual scenario where present processes and human influenced coastal activities continue unabated (FME, 2010).

Cost of coastal protection depends on the type of protection, the labour cost, the cost of planning, cost of engineering, equipment and materials; these however are influenced by the economic development level of each country (Hillen, et al., 2010). To put in context how much might be needed to protect the Niger delta, the cost of implementing the adaptation measures (simulated in this chapter) in countries around the world are given in table 11.

Table 11. Example costs of the proposed coastal interventions as applied in different places around the world.

No	Protection type	Cost	Country	Reference
1	Dykes	€4-22 million/ km/m increase in height	Netherlands	(Hillen, et al., 2010)
		€4-22 million/ km/m increase in height	USA (New Orleans)	
2	Bypass channel	£300 - 400/m length (for a 1km channel)	UK	(Jones, Keating, & Pettit, 2015)
3	Dredging	£ 12 - 15 /m^3	UK	(Jones, Keating, & Pettit, 2015)
		$ 490/m	Nigeria	(Emiedafe, 2015)
4	Storm surge barrier	€450 million	Netherlands/ Maeslantkering	(Structurae, 1997) https://structurae.net/structures/maeslant-barrier
		€0.5–2.7 million / metre width	Global average	(Hillen, et al., 2010)

5	Coastline shortening	€4-22 million/ km/metre increase	Netherlands	(Hillen, et al., 2010)
6	Flooding of low-lying areas	Depends on the value of the land.	-	-
7	Doing nothing	-	-	-

Applying these costs to the area modelled here (depending on the intervention applied), the cost will be as follows:

- Use of dykes (total length 75.4m, as shown in figure 17.5) at €4 -22 million/km/m = €301 – 1,658.8 million/m height
- Use of a bypass channel (5km long. Figure 17.19) at £300 - 400/m length = £15 -20 million
- Construction of a storm surge barrier to cover 3km wide channel (figure 17.16) at €0.5– 2.7 million / metre width = € 1,500 – 8,100 million
- Coastline shortening using a 1.3km wide, 6m high 'dyke' at €4 -22 million /km/m = € 31,200 – 171,600 million
- Dredging is not recommended as an option to be used on this study area segment.

7.4 Conclusion

Adaptation strategies that are accepted and implemented depend on the value of the coastal area, the political and local considerations and the cost of the adaptation measures to be taken. These factors affect the type and extent of the adaptation plan to be applied; thus an adaptation plan can extend to all vulnerable areas, or be limited to economically viable areas and/or human settlements. Adaptation plans can have different aims/goals, criteria and modes of implementation.

Based on results of models run in chapters 5, the coastal vulnerability to SLR index computed in chapter 6, and the assessment of resilience in chapter 7, two possible SLR mitigation and adaptation scenarios were simulated for the Niger delta. The first scenario aimed at planning resilient settlements used four options: application of dykes around all vulnerable settlements in the modelling domain, use of bypass channel to divert flood waters, river dredging, and allowing inundation of low lying areas. The results of these interventions showed that with the exception of channel dredging, a combination of interventions will be effective against SLR up to 1.5m. As no inundation is allowed in the second scenario more interventions were added to stop any possibility of flooding. Thus interventions included use of a storm surge barrier, and coastline shortening. The results showed effective resistance to flooding, but shortening of the coastline also dried up inner channels which support many fishing communities.

Although not simulated in this chapter, another intervention that will be good for the Niger delta environment is restoration of wetlands via replanting of degraded Mangrove; the effectiveness of such has been shown by locals planting Bamboo/ plantain as erosion mitigation options (chapter 6).

Finally, an integrated approach to coastal management that improves protection of coastal environment while creating a good habitat for natural and human activities, is the way forward for the Niger delta.

8

Conclusions and Recommendations

The general aim of this thesis was to assess the impact of Sea Level Rise (SLR) on the Niger delta land area, coastline, and surface water in an integrated way that will lead to practical recommendations for adaptation.

8.1 Conclusion

SLR impact studies require a methodology that assesses the coastal area in such a way that its characteristics are studied, the important factors affecting it are assessed, and its future needs are documented (IPCC, 1992). In accordance with this, this study combined hydrodynamic modelling, GIS analysis, and vulnerability index computations to fulfil its objectives and answer the research questions as elaborated in the subsequent subsections.

8.1.1 How can satellite data be applied in hydrological studies in delta areas?

Satellite remote sensing has been applied in hydrology for many years, with the earliest applications in water body and flood mapping (chapter 3). Beyond mapping, satellite data in the form of imagery, DEM, altimetry data, etc., can be used as hydraulic model input forcing factors or to constrain model data during calibration/validation/verification (Pereira-Cardenal, et al., 2011). Satellite-based estimates of river flow, river width, water levels and flooding extent are used for model calibration/validation/ and verification. Although data quality, pre/post data processing etc. introduce errors in satellite derived data and increase the uncertainties in satellite data utilization, several methods have been developed to quantify these errors and produce acceptable results.

As a data scarce area, studies on the Niger delta have been limited by data availability. Consequently there exists a gap in the methodology for quantitative and qualitative analysis of the implications of SLR for the Niger delta. This thesis used satellite data to estimate river cross sections, classify and determine roughness values of land-use and land-cover, calculate river width, as input into hydrodynamic models and for model calibration. Satellite derived information were also used; these include flood maps, settlement size and location, distance between features and tidal water levels. Water level data obtained during river dredging campaigns provided another non-conventional data source used in this thesis.

8.1.2   With recent increase in flooding, will sea level rise exacerbate the effects of river flooding? What is the effect on surface water?

In this thesis we simulated the interaction between the strong hydraulic currents flowing through the fresh water zone, and the flooding it causes in the coastal zone using the Sobek hydrodynamic 1D/2D model (chapter 4). The modelling results with SLR and SLR plus subsidence were compared with no SLR scenarios and checked for increase in flooding extent, extension of flooding time and change in water depth.

The results indicate that flooding in the lower Niger River will be affected by rise in sea levels especially as the area continues to subside. The effects include earlier occurrences of downstream flooding, increase in water depth and flooding of areas further upstream (than would occur without SLR). Although lateral flooding extent might not expand in the downstream areas, flooded areas will increase upstream because higher sea levels downstream will impede downward flow of flood waters which can result in a backwater effect and subsequent flooding of areas upstream. For years with no flooding from upstream, SLR will cause coastal areas downstream of the Niger River to flood earlier than usual. Moreover areas upstream of the Nun River which remain dry in normal years, will get flooded when sea levels rise.

8.1.3   What is the effect of sea level rise on coastal flooding and inundation?

This thesis modelled the effect of coastal flooding using Deltares' DFlow modelling suite, which uses a flexible mesh for discretization of the modelled area (Chapter 4). The Bonny and New Calabar river system located on the eastern Niger delta was modelled using an unstructured grid. The use of an unstructured mesh for hydrodynamic modelling allowed for finer grid spacing around areas where river flow dynamics change more rapidly (such as bends) and it also the allowed use of a triangular grid type (with coarser grid spacing) on the floodplain. The model outputs were analysed for flooding and inundation, increase in land loss, water depth and flood extent as compared with no sea level rise scenarios.

Model results indicated that SLR will increase the occurrence of coastal flooding (this is indicated by the flood generated by even the lowest level of rise in sea levels) because water levels and water depths in Bonny River will be higher, thus increasing land area flooding extent. The flow velocity is also much higher with SLR than without. Coastal flood waters will thus be transported faster along the river to places upstream. Consequently, flooding of land areas at high tide will increase due to higher water levels. Furthermore, differences in flow velocity around narrow bends will also be higher with SLR than without SLR; making river crossings more dangerous.

8.1.4   How much of the land can be lost to inundation?

GIS-based 'Bathtub approach' was used to compute possible inundation extents from future rise in sea levels using SRTM DEM, a tidal surface, SLR and land subsidence values (chapter 6). Using projected global eustatic SLR values of 19mm by 2030 and 35mm by 2050 in addition to subsidence, RSLR levels for the Niger delta to range from 0.14m - 0.44m by 2030, and 0.29m - 0.96m by 2050. Since the elevation of an area determines the lowest level of water that can flood it, the bathtub approach subtracts the topography from the sum of tidal water levels and SLR to determine areas likely to be inundated.

The results showed that a rise in sea levels of 0.14m already inundates areas in the Niger delta, and the flood extent increases with increase in SLR. Also, inundated water depths increased with higher SLR values; thus areas already inundated at lower SLR are likely to have deeper water depths with increase in sea levels. Consequently, 4.6-5.2% (1119.3 -1254km^2) of Niger delta land area can be lost to inundation by 2030, and 4.9-6.8% (1175.9 -1633km^2) by 2050.

In conclusion, with eustatic SLR predicted to reach 1m by 2100 and continued subsidence in the Niger delta due to oil and gas exploration, there is high probability of further inundation and loss of land. Furthermore, the results indicate that without subsidence the inundation effect of SLR on the Niger delta will be minimal (since the eustatic values are just 19mm and 35mm by 2030 and 2050 respectively). Subsidence has therefore made the Niger delta very vulnerable to inundation by making the RSLR values very high.

8.1.5   How can the vulnerability of deltaic coastlines to sea level rise be evaluated?

Highly vulnerable coastlines expose the inland areas to effects of SLR, serving as a gateway for inundation, storm surge and coastal erosion. This thesis developed a new coastal vulnerability index called coastal vulnerability index due to SLR ($CV_{SLR}I$) which evaluates coastline vulnerability to SLR using 17 physical, social and human influence indicators of exposure, susceptibility and resilience (Chapter 5). The results indicated that 42.6% of the Niger delta is highly vulnerable to effects of SLR like flooding, erosion, and salt water intrusion into underground aquifers. Such vulnerable areas are characterized by low slope (<1%), low topography (3-5m), estuaries, very high hydraulic conductivity (>81m/day), very high population density (>800 people/km^2), and settlements within 100-200m of the coastline. Besides, analysis of the social and human influence variables showed that in terms of aquifer type, aquifer hydraulic conductivity, population growth, sediment supply, and groundwater consumption, the Niger delta is vulnerable to effects of SLR. The combination of these properties has made the coastal segments highly vulnerable to SLR.

In conclusion, this thesis shows the variables that make a coastline highly vulnerable to SLR include physical coastal properties, human influence, and social properties. The presence of human influence variables such as coastal infrastructure and high population density, increase the probability of damage to lives and property when a disaster occurs. Human influence on coastal environments can affect sediment supply and accelerate erosion, and should therefore be captured in vulnerability assessments. Also, the location of many settlements in remote areas, far away from the local government headquarters, reduces the value of resilience to effects of SLR.

8.1.6   What should be considered in planning for SLR adaptation? Are there existing sustainable options that can be used?

Adaptation planning requires careful consideration of the social and physical properties of an area, as well as the resilience of the inhabitants to effects of SLR (chapter 6). In this thesis physical and social coastal properties analysed include slope, topography, aquifer /borehole conditions, settlement proximity to the coastline, and population density. The results show that the Niger delta is vulnerable to further flooding, inundation, intrusion of sea salts, and erosion from SLR. In terms of resilience, local response strategies used by the people and the degree of effectiveness of the methods in enabling them to cope were reviewed and analysed for their suitability as future adaptation strategies. The review showed that local practices have helped people to cope with the challenges posed by flooding, erosion, inundation, and inland intrusion of sea salts. However some of the practices have disadvantages that make them undesirable for inclusion in future planning.

In conclusion, sustainable local practices in the Niger delta include: planting of Bamboo trees for erosion control, use of sandbags as bridges and dykes (flood control), use of flood receptor pits as temporary flood water reservoirs, and community legislation against sand mining and indiscriminate tree felling.

8.1.7 What are the possible effects of mitigation/adaptation options on SLR impacts on the Niger delta?

Two possible SLR mitigation and adaptation scenarios were simulated (chapter 7) for the Niger delta. The first scenario aimed at planning resilient settlements, used four options:(i) application of dykes around all vulnerable settlements in the modelling domain, (ii) use of bypass channel to divert flood waters, (iii) river dredging, and (iv) allowing inundation of low lying areas. The results of these interventions showed that with the exception of channel dredging, a combination of interventions will be most effective against SLR up to 1.5m. As no inundation is allowed in the second scenario, more interventions were added to stop any possibility of flooding. The extra interventions included use of a storm surge barrier, and coastline shortening. The results showed effective resistance to flooding, but shortening of the coastline also dried up inner channels which support many fishing communities.

The conclusion from chapters 6 and 7 therefore is that suitable adaptation measures for the Niger delta include: construction of dykes, by-pass channels, flood-pits (reservoirs), storm surge barrier, and coastline shortening. To effectively adjust to living with SLR on the Niger delta the following strategies should be adopted: building new structures raised above ground, change of farming practices, and legislation to ensure compliance by all. Other practices that can be effective especially around beach covered coastlines, include: dune sand nourishment, construction of offshore groins, sea walls, and barrier revetment.

8.2 Recommendations

1   Although in this study simulations and analyses were done for years 2030 and 2050 based on projected SLR values for 2100, the reality is that oil and gas exploration continues and so subsidence continues. SLR values for the Niger delta are most likely to exceed the modest values used here. Therefore just like the recommendation by Williams & Ismail (2015) that SLR global projections should extend to 2 m in addition to local geophysical and man-made factors, any coastal protection structures built should be designed such that they can be extended to protect against higher SLR values by 2100.

2   In order to obtain better results with modelling approaches it is recommended that reliable subsidence values in the Niger delta be obtained via measurement surveys undertaken at diverse sites. Such subsidence rates when obtained can show the rates at sites undergoing oil and gas extraction and those undergoing natural subsidence, thus enabling better models of effects of subsidence can be developed for the Niger delta.

3   In chapter 6, we highlighted some local practices that are suitable to combat flooding, inundation and erosion due to sea level rise in the Niger delta. However detailed studies and measurements of the effectiveness of the local practices should be undertaken to determine the limit of their effectiveness and how to precisely improve them.

4   Since this study was limited to onshore areas, the mitigation/adaptation measures were only applied onshore. For complete and effective SLR adaptation planning however, there might be a need to include offshore measures like Seawalls, Groins/ Breakwaters (measures that can interrupt and change flow direction or amplitude before it reaches the coastline). This is particularly important for sandy coastlines. It is recommended that studies on such offshore interventions should also be undertaken.

5   The Niger delta presently lacks effective gauging stations where hydrological measurements can be taken. To enable a practical cost estimation, a database of factors related to SLR should be built. Such factors include data on rivers (water levels, flows, bathymetry, and water quality), land (topography, land-use, land-cover), coastal estuaries (water quality, water levels/flows) wetlands (area, biodiversity, and vegetation), nearshore (area, water levels/ tidal changes, land-use etc.). Therefore gauging and survey stations should be installed in different locations within the Niger

delta in order to take these measurements. Satellite data on the Niger delta should also be obtained from all sources possible.

6   In chapter 6 and 7 of this thesis, the need for laws that will protect the vulnerable Niger delta areas and enable resilience and adaptation, are mentioned. These laws include those against indiscriminate sand mining, deforestation, and building/expansion of settlement to areas identified as low-lying and vulnerable to flooding/inundation. In terms of government policies and laws, there are many within the Nigerian Environmental, Agricultural and Heath ministries that relate to hazards and disaster management. However, these consist of broad policies, and therefore need to be streamlined and adapted for SLR response strategies. That way, resilience issues can be effectively documented, improved and encouraged among local people. Particularly, in order to avoid confusion and duplication of schedules, the hierarchy of implementers of SLR strategies should be made clear between the Ministry of Environment, the Ministry of the Niger delta, the various state governments and agencies, and the Niger Delta Development Commission (NDDC).

8.3 Study Limitations

This study was based on a number of assumptions as stated in each chapter. These assumptions introduce conditions that affect the accuracy of parts of the methods applied. Some of the limitations of this study include:

- Although effects of possible adaptation measures are simulated in chapter 7, no attempt has been made to design or compute the exact dimensions of structures used (e.g. dykes, surge barriers). Dimensions of structures used are examples only.

- The mapping of vulnerability as presented in the study is within limited bounds of the data accuracy and the scale of the study. Even though the local data used is acquired from official sources (see Table 4), there might still be uncertainties in the data collection and methods of processing that cannot be accounted for, because officially data is accepted as reliable by the authorities in charge of managing the delta.

- The classification of resilience based on communication penetration depended on indirect associations like settlement size and location and not on sufficient information about the communication type.

- The study did cover inland intrusion of sea water into ground water sources, but only considered that inundated areas will have sea water intrusion into surface water.

- This study used two uniform subsidence values across the Niger delta (7mm/year for lower subsidence and 25mm/year for high subsidence).

- The discharge capacity of 40-100m^3/s used for the simulated bypass channel (in chapter 7) was estimated from the model results.

- The present study is limited to onshore areas and does not include the vulnerability of offshore areas or their mitigation/adaption to SLR options.

Appendices

A. Comparison of SRTM data accuracy with 40m topographic contour data of Nigeria (chapter 3)

A 40m interval contour data of Nigeria was used to check the vertical accuracy of SRTM DEM data around the study area; both SRTM 1 arc second and 3 arc second DEMs. The contour data used lies right of the study area as shown in figure 0.1; since the study area covers a river and its floodplain that are generally below 40m, only a small part of the available contour data covers it, thus the need to use SRTM DEM for modelling. The contour data, obtained from the Geological surveys of Nigeria was converted into a Raster DEM and resampled to have same resolution as the SRTM DEMs; ArcGIS map algebra was used to calculate the residuals between the raster DEM and the SRTM DEMs; which was used to calculate the root mean square errors (RMSE).

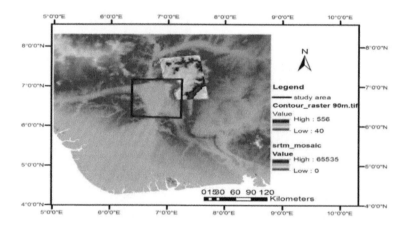

Figure 0.1 showing area with available contours used to compare the DEM. The low-lying floodplain area within the box is not covered by the 40m contour data of Nigeria.

The RMSE results for the lowest lying areas (40m) as shown in figure 0. 2(d) can give an error as low as 2.2m with SRTM 90m, and 2.6m for SRTM 30m for the low lying areas. At higher elevations the vertical errors of SRTM DEM show much higher values on high lands (where there is abrupt change in slopes) than low lying areas. This is exemplified in figure 0.2(c) where at contour elevation 120m, the RMSE of SRTM 30 is 6.55 and SRTM 90 is 5.42; this is very high compared to figure 0.2(d) with 40m contour elevation and RMSE 2.6 and 2.1 for SRTM 30m and 90m respectively.

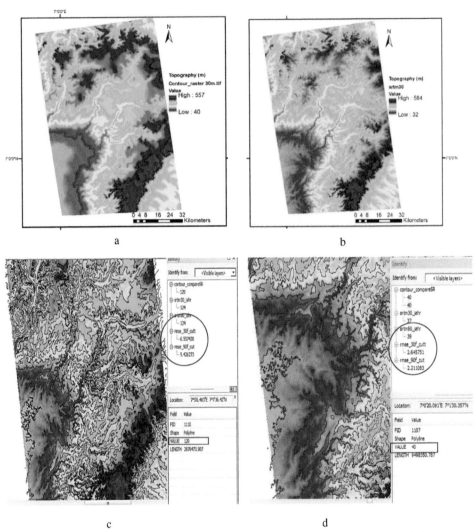

c   d

Figure 0.2. (a) - Rasterized 40m contour data around the study area. A.0.2 (b) SRTM 30m DEM of same area. A.0.2(c) and A.0.2 (d) examples of RMS error range with elevation. Low lying areas show less error.

B. Results of river flood modelling for Forcados and Nun rivers (chapter 4)

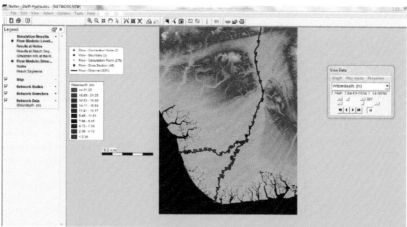

Figure 0.3. 1D Sobek model simulation results for 2006 flooding on Forcados and Nun River by 04/10/2006. Colour changes from blue to red with increase in water depth.

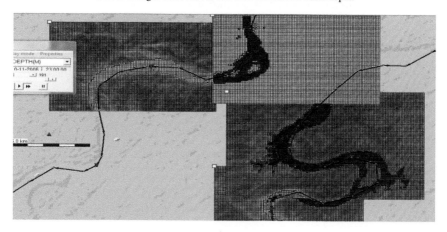

Figure 0.4. 1D/2D Sobek model simulation results showing flooded areas in blue. 90x90m SRTM DEM cover floodplain areas to show flooding in 2D.

References

Abam, T. (2001). Regional hydroiogical research perspectives in the Niger Delta. *Hydrological Sciences, 46*(1). Retrieved from http://hydrologie.org/hsj/460/hysj_46_01_0013.pdf

Adaramola, M., Oyewola, M., Ohunakin, O., & Akinnawonu, O. (2014). Performance evaluation of wind turbines for energy generation in Niger Delta, Nigeria. *Sustainable Energy Technologies and Assessments, 6*(2014), 75-85. doi:dx.doi.org/10.1016/j.seta.2014.01.001

Agabi, C. (2013). *Nigeria: 40km of Niger-Delta Land May Be Extinct in 20 Years – Don.* Retrieved April 29th, 2016, from http://mangroveactionproject.org/nigeria-40km-of-niger-delta-land-may-be-extinct-in-20-years-don/

Aich, V., Liersch, S., Vetter, T., Huang, S., Tecklenburg, J., Hoffmann, P., . . . Hattermann, F. (2014). Comparing impacts of climate change on streamflow in four large African river basins. *Hydrology and Earth System Sciences, 18*, 1305-1321. doi:doi:10.5194/hess-18-1305-2014, 2014.

Akinro, a., Opeyemi, D., & Ologunaba, I. (2008). Climate change and environmental degradation in the Riger delta region of nigeria: its vulnerability, impacts and possible mitigation. *Research Journal of Applied Sciences, 3*(3), 167- 173.

All science fair projects. (2015). *Sea water and its effects on Bambusa vulgaris (common bamboo).* Retrieved January 07, 2016, from http://www.all-science-fair-projects.com/print_project_1073_120

Alwang, J., Siegel, P., & Jorgensen, S. (2001). *Vulnerability a view from different disciplines.* Washington DC: World Bank. Retrieved from http://documents.worldbank.org/curated/en/2001/06/1637776/vulnerability-view-different-disciplines

Appiah, S. (2016). *Prez Mahama directs extension of Keta sea defence project to cover more communities.* Retrieved February 20, 2017, from Graphic Online: http://www.graphic.com.gh/news/general-news/prez-mahama-directs-extension-of-keta-sea-defence-project-to-cover-more-communities.html

Awosika, L., & Folorunsho, R. (2012). *Nigeria.* Retrieved from ODINAFRICA: http://www.odinafrica.org/nigeria

Awosika, L., French, G., Nicolls, R., & Ibe, C. (1992). The impact of sea level rise on the coastline of Nigeria. *IPCC Symposium on Global climate change and the Rising Challenges of the Sea* (p.

690 pp). Margarita, Venezuela: National Oceanic and Atmospheric Administration. Retrieved 2012

Bachmair, F., Calderon, S., Clarke, K., Davies, A., Davaadorj, K., Jenkins, P., . . . Shearer, K. (2012). *Climate Change Adaptation Capstone workshop .Concept paper: The Niger Delta. Spring 2012.* Columbia, Uk: Columbia SIPA.

Bacon, P. (1996). Wetlands and Biodiversity. In A. Hails (Ed.), *Wetlands, biodiversity and the Ramsar convention: the role of the convention on wetlands in the conservation and wise use of biodiversity* (p. 17). Gland: Ramsar Convention Bureau. Retrieved from Wetlands: cradle of species diversity .

Baird: Oceans, Lakes and Rivers. (2011). *Keta Coastal Defence.* Retrieved February 20, 2017, from http://www.baird.com/what-we-do/project/keta-coastal-defence

Balica, S. F., Douben, N., & Wright, N. G. (2009). Flood Vulnerability Indices at Different Spatial Scales. *water science and technology.*

Balica, S., Popescu, I., Beevers, L., & Wright, N. (2013). Parametric and physically based modelling techniques for flood risk and vulnerability assessment: a comparison. *Journal of Environmental Modelling and Software, 3*(41), 84-92.

Bariweni, P., Tawari, C., & Abowei , J. (2012). Some Environmental Effects of Flooding in the Niger Delta Region of Nigeria. *International Journal of Fisheries and Aquatic Sciences, 1*(1), 35-46.

Barneveld, H., Silander, J., Sane, M., & Malnes, E. (2008). Application of satellite data for improved flood forecasting and mapping. *4th International Symposium on Flood Defence:Managing Flood Risk, Reliability and Vulnerability Toronto, Ontario, Canada, May 6-8, 2008 .* Toronto, Canada. Retrieved May 31, 2015, from http://www.hkv.nl/site/hkv/upload/publication/Application_of_satellite_data_for_improved_f lood_forecasting_HJB.pdf

Belaud, G., Cassan, L., & Bader, J. C. (2010). Calibration of a propagation model in large river using satellite. In A. I. Stamou (Ed.), *6th International Symposium on Environmental Hydraulics* (pp. 869–874). Athens, Greece: CRC PRESS. Environmental Hydraulics.

Bercher1, N., Calmant, S., Picot, N., Seyler, F., & Fleury, S. (2014, 04 03). *Evaluation of cryosat-2 measurements for the monitoring of large river water levels.* Retrieved 06 25, 2015, from Along-Track.com: http://chronos.altihydrolab.fr/2012-09-23%20Venice%20ESA%2020%20years%20of%20progress%20in%20altimetry/Bercher.201 2b%20(Venice%20Paper)%20CryoSat-2%20hydro.pdf

Bhattacharya, B., Shams, M., & Popescu, I. (2013). On the influence of bed forms on flood levels. *Environmental Engineering Management, 12*, 857–863.

Bierbaum, R., Smith, J., Lee, A., Blair &, M., Carter, L., Stuart Chapin III &, F., . . . Verduzco, L. (2013). A comprehensive review of climate adaptation in the United States: more than before,but less than needed. *Mitigation Adaptation Strategy Global Change, 18*, 361-406. doi: 10.1007/s11027-012-9423-1

Birkmann, J. (2007). Risk and vulnerability indicators at different scales:Applicability, usefulness and policy implications. *Environmental Hazards*, 20-31.

Bogardi, J., Villagrán, J., Birkmann, J., & Renaud, F. (2005). Vulnerability in the context of Climate Change. *Human security and climate change*, (p. 14). Asker near Oslo. Retrieved January 12, 2016

Brakenridge, G., Kettner, A., Slayback, D., & Policelli, F. (2007, 01 23). *The Surface Water data record: Dartmouth Flood Observatory*. Retrieved from Dartmouth Flood Observatory: http://floodobservatory.colorado.edu/Version3/000E010Nv3.html

Briggs, N., Okowa, W., & Ndimele, O.-M. (2013). *The economic development of Rivers State*. Port Harcourt: River State economic advisory council. Retrieved August 17, 2015

Brooks, N. (2003). *Vulnerability, risk and adaptation: A conceptual framework*. Norwich, UK: Tyndall Centre for Climate Change Research. Retrieved from https://pdfs.semanticscholar.org/750a/8ad7921ba82db2f6395d3dec379355ac45cf.pdf

Brooks, N., Nicolls, R., & Hall, J. (2006). *Sea Level Rise Coastal Impacts and Responses*. Retrieved 09 18, 2012, from WBGU: www.wbgu.de/wbgu_sn2006.html

Brown, S., Kebede, A., & Nicolls, R. (2011). *Sea-Level Rise and Impacts in Africa, 2000 to 2100*. UNEP. Retrieved November 16, 2015, from http://www.unep.org/climatechange/adaptation/Portals/133/documents/AdaptCost/9%20Sea%20Level%20Rise%20Report%20Jan%202010.pdf

Buytaert, W., Dewul, A., De Bièvre, B., Clark, J., & Hannah, D. (2016). Citizen Science for Water Resources Management: Toward Polycentric Monitoring and Governance? *Journal of Water Resources Planning and Management*, 01816002-1-4. doi:DOI: 10.1061/(ASCE)WR.1943-5452.0000641

Buytaert, W., Zulkafli, Z., Grainger, C., Acosta, L., Alemie, T., Bastiaensen, J., . . . Zhumanova, M. (2014). Citeizen science in hydrology and water resources: opportunities for knowledge generation, ecosystem service management, and sustainable development. *Frontiers in Earth Science, 2*(26), 1-21. doi:doi: 10.3389/feart.2014.00026

Canadian Space Agency (CSA). (2015, 05 21). *RADARSAT Constellation.* Retrieved June 23, 2015, from RADARSAT constellation: http://www.asc-csa.gc.ca/eng/satellites/radarsat/

Cardno, C. (2016). *Massive flood protection project under way in bangladesh.* Retrieved January 29, 2017, from Civil Engineering: The Magazine Of The American Society Of Civil Engineers : http://www.asce.org/magazine/20150203-massive-flood-protection-project-under-way-in-bangladesh/

Castro-Gama, M., Popescu, I., Li, S., & Mynett, A. (2014). Flood inference simulation using surrogate modelling for the Yellow River multiple reservoir system. *Environmental Modelling and Software, 55*, 250–265.

Chen, H., Yang, D., Hong, Y., Gourley, J., & Zhang, Y. (2013). Hydrological data assimilation with the Ensemble Square-Root-Filter:Use of streamflow observations to update model states for real-time flash flood forecasting. *Advances in Water Resources, 59*, 209–220. doi:10.1016/j.advwatres.2013.06.010

Chow, V. T. (1959). *Open Channel Hydraulics.* Singapre: mcgraw-Hill, international edition 1973.

CIESIN. (2013, Januray 4). *Sea Level Rise Impacts on Ramsar Wetlands of International Importance.* (NASA Socioeconomic Data and Applications Center (SEDAC)) Retrieved January 13, 2016, from State of the planet: Earth institute,Center for International Earth Science Information Network, Columbia University. columbia University: http://dx.doi.org/10.7927/H4CC0XMD.

CNES. (2016). *AVISO+.* Retrieved November 22, 2016, from http://www.aviso.altimetry.fr/en/missions/future-missions/jason-cs.html

Crétaux, J. F., Stéphane, C., Romanovski, V., Shabunin, A., Lyard, F., Bergé Nguyen, M., . . . Perosanz, F. (2009). An absolute calibration site for radar altimeters in the continental domain : lake Issykkul in Central Asia. *Journal of Geodesy, 83*(8), 723-735.

Crétaux, J., Bergé-Nguyen, M., Leblanc, M., Del Rio, R. A., Delclaux, F., Mognard, N., . . . Maisongrande, P. (2011). Flood mapping inferred from remote sensing data. *Fifteenth International Water Technology Conference.* Alexandria, Egypt.

Cutter, S. L., Emrich, C. T., Webb, J. J., & Morath, D. (2009). *Social Vulnerability to ClimateVariability Hazards:A Review of the Literature.* South Carolina, USA: Hazards and vulnerability research institute. Retrieved from http://adapt.oxfamamerica.org/resources/Literature_Review.pdf

Cutter, S., Barnes, L., Berry, M., Burton, C., Elijah , E., Evans, E., . . . Webb, J. (2008). A place-based model for understanding community resilience to natural disasters. *Global Environmental Change, 18*(4), 598–606. doi:doi.org/10.1016/j.gloenvcha.2008.07.013

Danquah , J., Attippoe , J.-A., & Ankrah , J. (2014). Assessment of residential satisfaction in the resettlement towns of the keta basin in Ghana. *International Journal Civil Engineering, Construction and Estate Management, 2*(3), 26-45. Retrieved from http://www.eajournals.org/wp-content/uploads/Assessment-Of-Residential-Satisfaction-In-The-Resettlement-Towns-Of-The-Keta-Basin-In-Ghana.pdf

Dasgupta, e. a. (2009). *Climate Change and the future impacts of Storm Surge disasters in developing countries.* Centre for global dvelopment.

Dellepiane, S., de Laurentiis, R., & Giordano, F. (2004). Coastline Extraction from SAR Images and a Method for the Evaluation of the Coastline Precision. *Pattern Recognition Letters, 25,* 1461–1470.

Deltares. (2013, April). Sobek 1D/2D modelling suite for integral water solutions. *User manual*, pp. 491-500.

Deltares. (2016, 01 22). *D-Flow user manual.* Retrieved from http://content.oss.deltares.nl/Delft3d/manuals/D-Flow_FM_User_Manual.pdf

Di Baldassarre, G., Schumann, G., & Bates, P. D. (2009). A technique for the calibration of hydraulic models using uncertain satellite observations of flood extent. *Journal of Hydrology, 367*(3-4), 276–282. doi:10.1016/j.jhydrol.2009.01.020

Dinh, Q., Balica, S., Popescu, I., & Jonoski, A. (2012). Climate change impact on flood hazard, vulnerability and risk of the Long Xuyen Quadrangle in the Mekong Delta. *International Journal of River Basin Management, 10*(1), 103-120. doi:dx.doi.org/10.1080/15715124.2012.663383

DLR. (2015). *Earth Observation: TanDEM-X - the Earth in three dimensions.* Retrieved June 01, 2015, from http://www.dlr.de/dlr/en/desktopdefault.aspx/tabid-10378/566_read-426/#/gallery/345

DPIPWE. (2012). *Confined and Unconfined aquifers.* Retrieved 03 01, 2013, from http://www.dpipwe.tas.gov.au/inter.nsf/WebPages/JMUY-4Z5262?open

Dronkers, J., Gilbert, J., Butler, L., Carey, J., Campbell, J., James, E., . . . von Dadelszen, J. (1990). *Strategies for adaption to sea level rise. Report of the IPCC Coastal Zone Management Subgroup: Intergovernmental Panel on Climate Change.* Geneva: IPCC. Retrieved from http://papers.risingsea.net/federal_reports/IPCC-1990-adaption-to-sea-level-rise.pdf

Duan, Z., & Bastiaanssen, W. (2013). Estimating water volume variations in lakes and reservoirs from four operational satellite altimetry databases and satellite imagery data. *Remote Sensing of Environment, 134,* 403–416. doi:10.1016/j.rse.2013.03.010

Dwarakish, G. S., Vinay, S. A., Natesan, U., Asano, T., Kakinuma, T., Venkataramana, K., . . . Babita, M. K. (2009). Coastal vulnerability assessment of the future sea level rise in Udupi coastal zone of Karnataka state,west coast of India. *Ocean and Coastal management, 52*, 467-478. doi:http://dx.doi.org/10.1016/j.ocecoaman.2009.07.007

Ehiorobo, J., & Izinyon, O. (2011). Measurements and Documentation for Flood and Erosion Monitoring and Control in the Niger Delta States of Nigeria. *FIG Working Week 2011.* Marrakech, Morocco: FIG Working Week 2011. Retrieved 08 08, 2016

El-Rabbany, A. (2002). *Introduction to GPS: the global positioning system.* Norwood MA: ARTECH HOUSE INC.

Emiedafe, W. (2015). *8 Mega Construction Projects in Nigeria.* Retrieved March 13, 2017, from http://sapientvendors.com.ng/8-mega-construction-projects-in-nigeria/

Epete, B. (2012). The flow pattern of the bonny and new calabar river systems of niger delta region, Nigeria. FIG.NET. Retrieved from FIG Working Week 2012 – Territory, environment, and cultural heritage: https://www.fig.net/resources/proceedings/fig_proceedings

/fig2012/ppt/ts07g/TS07G_ekpete_5516_ppt.pdf

Ericson, J., Vorosmarty, C., Dingman, S., Ward, L., & Meybeck, M. (2006). Effective sea level rise and deltas: causes of change and human dimension implications. *Journal of Planetary Change*, 63-82.

ESA. (2012). *Space in Images: mean sea level trends.* Retrieved November 16, 2016, from European Space Agency: http://www.esa.int/spaceinimages/Images/2012/09/Mean_sea_level_trends

ESA. (2015). *Sentinel 3.* Retrieved June 15, 2015, from Copernicus, observing the Earth: http://www.esa.int/Our_Activities/Observing_the_Earth/Copernicus/Sentinel-3

ESRI. (2016). *ArcGIS.* Retrieved 01 06, 2016, from http://www.esri.com/software/arcgis/capabilities

European commission. (2012). *Integrated Coastal Zone Management. Our coast: outcomes and lessons learnt* (2012 ed.). (R. Steijn, P. Czerniak, A. Volckaert, M. Ferreira, E. Devilee, T. Huizer , & R. ter Hofstede, Trans.) Luxembourg: Luxembourg : Publications Office of the European Union. doi:doi : 10.2779/9345

Eyers, R., & Obowu, C. (2013). Niger Delta Flooding: Monitoring, Forecasting & Emergency Response Support from SPDC. Abuja, Nigeria, 6 – 10 May 2013: FIG Working week 2013. Retrieved 08 16, 2016

Ezer, T., & Liu, H. (2010). On the dynamics and morphology of extensive tidal mudflats: Integrating remote sensing data with an inundation model of Cook Inlet, Alaska. *Ocean Dynamics, 60*(5), 1307-1318. doi:10.1007/s10236-010-0319-x

FME. (2010). *National Environmental, Economic and Development Study (NEEDS) for climate change in Nigeria.* Federal Ministry of Environment (Special climate change unit). Retrieved from https://unfccc.int/files/adaptation/application/pdf/nigerianeeds.pdf

Fohringer, J., Dransch, D., Kreibich, H., & Schröter, K. (2015). Social media as an information source for rapid flood inundation mapping. *Natural Hazards and Earth Systems Sciences (NHESS), 15\,* 2725–2738. doi:10.5194/nhess-15-2725-2015

Fraser, C. S., & Ravanbakhsh, M. (2011). Performance of DEM generation technologies in coastal environments. *7th International Symposium on Digital Earth.* Perth, Australia.

French, T. (2016). *About Tech: What is Microsoft Excel and What Would I Use it for?* Retrieved 07 17, 2016, from http://spreadsheets.about.com/od/excelformulas/ss/What-is-Microsoft-Excel-and-What-Would-I-Use-it-for.htm

Fu , C., Popescu, I., Wang, C., & Mynett, A. (2014). Challenges in modelling river flow and ice regime on the Ningxia–Inner Mongolia reach of the Yellow River, China. *Hydrology and Earth Systems Sciences, 18,* 1225–1237. doi:doi:10.5194/hess-18-1225-2014

García-Pintado, J., Mason, D., Dance, S., Cloke, H., Neal, J., Freer, J., & Bates, P. (2015). Satellite-supported flood forecasting in river networks: A real case study. *Journal of Hydrology, 523,* 706–724. doi:10.1016/j.jhydrol.2015.01.084

Getirana, A. C., Bonnet, M.-P., Calmant, S., Roux, E., Rotunno Filho, O. C., & Mansur, W. J. (2009). Hydrological monitoring of poorly gauged basins based on rainfall–runoff. *Journal of Hydrology, 379,* 205–219. doi:0.1016/j.jhydrol.2009.09.049

Giustarini, L., Matgen, P., Hostache, R., Montanari, M., Plaza, D., Pauwels, V., . . . Savenije, H. (2011). Assimilating SAR-derived water level data into a hydraulic model:a case study. *Hydrology and Earth System Sciences, 15,* 2349–2365. doi:10.5194/hess-15-2349-2011

Good, S., Mallia, D., Lin, J., & Bowen, G. (2014). Stable Isotope Analysis of Precipitation Samples Obtained via Crowdsourcing Reveals the SpatiotemporaL Evolution of Superstorm Sandy. *PLoS ONE, 9*(3), 1-7. doi:doi:10.1371/journal.pone.0091117

Gornitz, V., Couch, S., & Hartig, E. (2001). Impacts of sea level rise in the new york city metropolitan area. *Global and planetary change, 32*(1), 61-88.

Gornitz, V., White, T., & Cushman, R. (1991). Vulnerability of the U.S. to future sea-level rise. *Seventh Symposium on Coastal and Ocean Management,* (pp. 2354-2368). Long Beach, CA(USA).

Gorokhovich, Y., & Voustianiouk, A. (2006). Accuracy assessment of the processed SRTM-based elevation data by CGIAR using field data from USA and Thailand and its relation to the terrain characteristics. *Remote Sensing of Environment, 104*, 409–415. doi:10.1016/j.rse.2006.05.012

Hexagon Geospatial. (2016). *ERDAS IMAGINE*. Retrieved March 16, 2016, from http://www.hexagongeospatial.com/products/producer-suite/erdas-imagine

Hillen, M., Jonkman, S., Kanning, W., Kok, M., Geldenhuys, M., & Stive, M. (2010). *Coastal defence cost estimates: case stydy of the Netherlands, New Orleans and Vietnam*. Delft, Netherlands: TU Delft.

Horritt, M. S. (2006). A methodology for the validation of uncertain flood inundation models. *Journal of Hydrology, 326*(1-4), 153–165. doi:10.1016/j.jhydrol.2005.10.027

Ibe, A. (1988). *Coastline erosion in nigeria*. Ibadan, Nigeria: University press.

IPCC. (1992). *Intergovernmental panel on Climate Change Common Methodology*. The Hague, Netherlands: IPCC- CZMS.

IPCC. (2007). *Regional Impacts of Climate Change*. IPCC. Retrieved November 17, 2015, from https://www.ipcc.ch/ipccreports/sres/regional/503.htm

IPCC. (2007a). *Natural System responses to climate Change Drivers*. Working group 11: Impacts, Adaptation and Vulnerability. Inter Governmental panel on Climate Change. Retrieved from http://www.ipcc.ch/publications_and_data/ar4/wg1/en/ch5s5-5-2-2.html

IPCC. (2007b). *Deltas*. Working group 11: Impacts, Adaptation and Vulnerability. Retrieved from Intergovernmental panel on Climate Change.

IPCC. (2013). *Climate Change 2013: The Physical science Basis. Contribution of working group 1 to the Fifth Assessment Report of the Intergovernmental panel on Climate Change*. Cambridge, UK: Cambrige University press. From http://www.ipcc.ch/pdf/assessment-report/ar5/wg1/WG1AR5_Chapter13_FINAL.pdf

IPCC. (2013). *Working group I contribution to the IPCC fifth assessment report (AR5), climate change 2013: the physical science basis*. Stockholm: IPCC. Retrieved from http://www.climatechange2013.org/images/uploads/WGIAR5_WGI-12Doc2b_FinalDraft_Chapter13.pdf

Islam, R. Z., Begum, S. F., Yamaguchi, Y., & Ogawa, K. (2002). Distribution of suspended sediment in the coastal sea off the Ganges–Brahmaputra River mouth: observation from TM data. *Marine Systems, 32*(4), 307–321. doi:10.1016/S0924-7963(02)00045-3

Jarihani, A., Callow, J., Johansen, K., & Gouweleeuw, B. (2013). Evaluation of multiple satellite altimetry data for studying inland water bodies and river floods. *Journal of Hydrology, 505*, 78–90. doi:10.1016/j.jhydrol.2013.09.010

Jarihani, A., Callow, J., McVicar, T., Van Niel, T., & Larsen, J. (2015). Satellite-derived Digital Elevation Model (DEM) selection, preparation and correction for hydrodynamic modelling in large, low-gradient and data-sparse catchments. *Journal of Hydrology, 524*, 489–506. doi:10.1016/j.jhydrol.2015.02.049

Jarihani, A., McVicar, T., Van Niel, T., Emelyanova, I., Callow, J., & Johansen, K. (2014). Blending Landsat and MODIS Data to Generate Multispectral Indices: A Comparison of "Index-then-Blend" and "Blend-then-Index" Approaches. *Remote Sensing, 6*, 9213-9238. doi:10.3390/rs6109213

Jena, P., Panigrahi, B., & Chatterjee, C. (2016). Assessment of Cartosat-1 DEM for Modeling Floods in Data Scarce Regions. *Water Resources Management, 30*, 1293–1309. doi:DOI 10.1007/s11269-016-1226-9

Jones, P., Keating, K., & Pettit, A. (2015). *Cost estimation for channel management summary of evidence.* BrisTol: Environment Agency. Retrieved from http://evidence.environment-agency.gov.uk/FCERM/Libraries/FCERM_Project_Documents/SC080039_cost_channel_mg mt.sflb.ashx

Jonoski, A., & Popescu, I. (2012). Distance Learning in Support of Water Resources Management: An Online Course on Decision Support Systems in River Basin Management. *Water Resources Management, 26*(5), 1287–1305. doi:10.1007/s11269-011-9959-y

Jung, C. H., Hamski, J., Durand, M., Alsdorf, D., Hossain, F., Lee, H., . . . Hasan, K. (2010). Characterization of complex fluvial systems using remote sensing of spatial and temporal water level variations in the Amazon, Congo, and Brahmaputra Rivers. *Earth Surface Processes and Landforms, 35*(3), 294-304. doi:10.1002/esp.1914

Karim, M., & Mimura, N. (2008). Impacts of climate change and sea-level rise on cyclonic storm surge floods in Bangladesh. *Global Environmental Change, 18*(3), 490–500. doi:dx.doi.org/10.1016/j.gloenvcha.2008.05.002

Karlsson, J. M., & Arnberg, W. (2011). Quality analysis of SRTM and HYDRO1K: a case study of flood inundation in Mozambique. *International Journal of Remote Sensing, 32*(1), 267-285. doi:10.1080/01431160903464112

Khan, S. I., Hong, Y., Vergara, H. J., Gourley, J. J., Brakenridge, G. R., De Groeve, T., . . . Yong, B. (2012). Microwave Satllite Data for Hydrologic Modelling in Ungauged Basins. *IEEE Geoscience and Remote Sensing Letters, 9*(4). doi:0.1109/LGRS.2011.2177807

Kim, J., Lu, Z., Lee, H., Shum, C. K., Swarzenski, C. M., Doyle, T. W., & Baek, S.-H. (2009). Integrated analysis of PALSAR/Radarsat-1 InSAR and ENVISAT altimeter data for mapping of absolute water level changes in Louisiana wetlands. *Remote Sensing of Environment, 113*(11), 2356-2365.

Kiptala, J., Mul, M., Mohamed , Y., & Van der Zaag, P. (2014). Modelling stream flow and quantifying blue water using a modified STREAM model for a heterogeneous, highly utilized and data-scarce river basin in Africa. *Hydrology and Earth Sytem Scieneces Journal (HESS), 18*, 2287–2303. doi:10.5194/hess-18-2287-2014

KPN.nl. (2017). *Droogmakerij en Polderhistorie.* Retrieved February 23, 2017, from http://home.kpn.nl/keesbolle/HoofdHist.html

Kumar, A., Narayana, A., & Jayappa, K. (2010a). Shoreline changes and morphology of spits along southern Karnataka, west coast of India: A remote sensing and statistics-based approach. *Geomorphology, 120*, 133–152. doi:10.1016/j.geomorph.2010.02.023

Kumar, T., & Kunte, P. (2012). Coastal Vulnerability Assessment for Chennai, East coast of India using Geospatial Techniques. *Journal of Natural Hazards, 64*, 853-872.

Laukkonen, J., Blancob, P.-K., Lenhart, J., Keinerd, M., Cavrice, B., & Kinuthia-Njengaf, C. (2009). Combining climate change adaptation and mitigation measures at the local level. *Habitat International, 33*(3), 287–292. doi:10.1016/j.habitatint.2008.10.003

Leauthaud, C., Belaud, G., Duvail, S., & Moussa, R. (2013). Characterizing floods in the poorly gauged wetlands of the Tana River Delta, Kenya, using a water balance model and satellite data. *Hydrological Earth Systems Science, 17*, 3059–3075. doi:10.5194/hess-17-3059-2013

León, J. G., Calmant, S., Seyler, F., Bonnet, M.-P., Cauhopé, M., Frappart, F., . . . Fraizy, P. (2006). Rating curves and estimation of average water depth at the upper Negro. *Journal of Hydrology, 328*, 481-496. doi: 10.1016/J.JHYDROL.2005.12.006

Li, S., Sun, D., Goldberg, M., & Stefanidis, A. (2013). Derivation of 30-m-resolution water maps from TERRA/MODIS and SRTM. *Remote Sensing of Environment, 134*, 417–430. doi:10.1016/j.rse.2013.03.015

Liew, S. C., Gupta, A., Wong, P. P., & Kwoha, L. K. (2010). Recovery from a large tsunami mapped over time: The Aceh coast, Sumatra. *Geology, 114*(4), 520-529. doi:10.1016/j.geomorph.2009.08.010

Lillesand, T., Kiefer, R. W., & Chapman, J. W. (2004). *Remote sensing and image interpretation.* (5th, Ed.) John Wiley and sons.

Long, S., Fatoyinbo, T., & Policelli, F. (2014). Flood extent mapping for Namibia using change detection and thresholding with SAR. *Environmental Research Letters, 9*(035002). doi:10.1088/1748-9326/9/3/035002

Martin, V. N., Pires, R., & Cabral, P. (2012). Modelling of coastal vulnerability in the stretch between the beaches of Porto de Mós and Falésia,Algarve (Portugal). *Journal of Coastal Conservation*, 503-510. doi:DOI 10.1007/s11852-012-0191-6

Mason, D. C., Horritt, M. S., Dall'Amico, J. T., & Scott, T. R. (2007). Improving River Flood Extent Delineation From Synthetic Aperture Radar Using Airborne Laser Altimetry. *IEEE Transactions on Geoscience and Remote Sensing and*(0196-2892), 3932 - 3943. doi:10.1109/TGRS.2007.901032

Mason, D. C., Scott, T. R., & Dance, S. L. (2010). Remote sensing of intertidal morphological change in Morcambe Bay, U,K., between 1991 and 2007. *Estuarine , Coastal and Shelf Science, 87,* 487-496. doi:10.1016/j.ecss.2010.01.015

Matgen, P., Montanari, M., Hostache, R., Pfister, L., Hoffmann, L., Plaza, D., . . . Savenije, H. (2010). Towards the sequential assimilation of SAR-derived water stages into hydraulic models using the Particle Filter: proof of concept. *Hydrology and Earth System Sciences, 14*, 1773–1785. doi:10.5194/hess-14-1773-2010

Mclaughlin, S., & Cooper, J. A. (2010). A multi-scale coastal vulnerability index: A tool for coastal managers? *Environmental Hazards*, 233-248. doi:10.3763/ehaz.2010.0052

Mclaughlin, S., Mckenna, J., & Cooper, J. (2002). Socio-economic Data in Coastal Vulnerability Indices: Constraints and Opportunities. *Journal of coastal research*, 487-489.

Mcmanus, J. (2002). Deltaic responses to Changes in River Regimes. *marine Chemistry, 79*, 155 - 170.

Mcmillan, H., Hreinsson, E., Clark, M., Singh, S., Zammit, C., & Uddstrom, M. (2013). Operational hydrological data assimilation with the recursive ensemble Kalman filter. *Hydrology and Earth System Sciences, 17*, 21-38. doi:10.5194/hess-17-21-2013

Meijerink, A. M., Bannert, D., Batelaan, O., Lubczynski, M. W., & Pointet, T. (2007). *Remote sensing applictions to groundwater.* Series on Groundwater, UNESCO.

Michelsen, N., Dirks, H., Al-Saud, M., & Schüth, C. (2016). YouTube as a crowd-generated water level archive. *Science of the Total Environment, 568*(2016), 189–195. doi:http://dx.doi.org/10.1016/j.scitotenv.2016.05.211

Middlesex community College (MXCC). (2013). *What is a database?* Retrieved 09 09, 2016, from http://mxcc.edu/wp-content/uploads/2013/09/InternetVsDBs.pdf

Moya Quiroga, V., Popescu, I., & Solomatine, D. (2013). Cloud and cluster computing in uncertainty analysis of integrated flood models. *Journal of Hydroinformatics, 15*, 55–70.

Muis, S., Güneralp, B., Jongman, B., C.J.H. Aerts, J., & Ward, P. (2015). Flood risk and adaptation strategies under climate change and urban expansion: A probabilistic analysis using global data. *Science of The Total Environment, 538*, 445–457. doi:10.1016/j.scitotenv.2015.08.068

Musa, Z. N., Popescu, I., & Mynett, A. (2014b). Modelling the effects of sea level rise on flooding in the lower Niger River. *11 international conference on Hydro-informatics, HIC 2014.* New York: HIC, 2014. Retrieved from https://www.conftool.pro/hic2014/index.php?page=browseSessions&form_session=90

Musa, Z., Popescu, I., & Mynett, A. (2014a). Niger delta's vulnerability to river floods due to sea lavel rise. *Natural Hazards and Earth System Science (NHESS), 14*, 3317-3329. doi:10.5194/nhess-14-3317-2014

Musa, Z., Popescu, I., & Mynett, A. (2015). A review of applications of satellite SAR, optical,altimetry and DEM data for surface water modelling, mapping and parameter estimation. *Hydrology and Earth Sustems Sciences (HESS), 19*, 3755-3769. doi:10.5194/hess-19-3755-2015

Musa, Z., Popescu, I., & Mynett, A. (2016). Assessing the sustainability of local resilience practices against sea level riseimpacts on the lower Niger delta. *Ocean and Coastal management, 130*(2016), 221-228. doi:http://dx.doi.org/10.1016/j.ocecoaman.2016.06.016

Nairn, R., Macintosh, K., Hayes, M., Nai, G., Anthonio, S., & Valley, W. (1998). Coastal Erosion at Keta Lagoon, Ghana - Large Scale Solution to a Large Scale Problem. *Journal of Coastal Engineering*, 3192- 3205. Retrieved from https://icce-ojs-tamu.tdl.org/icce/index.php/icce/article/viewFile/5833/5501

NASA. (2012). *Flooding in Nigeria.* Retrieved 08 12, 2016, from http://earthobservatory.nasa.gov/IOTD/view.php?id=79404

NBCC: National Black Chamber of Commerce. (2011). *Sea Defense and Erosion Projects, Ghana.* Retrieved February 20, 2017, from https://www.nationalbcc.org/resources/contracting/1299-sea-defense-and-erosion-projects-ghana

NBS. (2011). Women in transforming Nigeria. *Gender statistics newsletter: Quarterly publication of the Nigerian bureau for statistics, 2*(4). Retrieved 12 12, 2016, from http://www.nigerianstat.gov.ng/report/39

NDDC. (2014). *NDDC, making a difference: NDDC boss clears Ibeno mega bridge for commissioning~promises more for niger delta.* Retrieved December 15, 2015, from http://www.nddc.gov.ng/news_id5s.html

NDRMP. (2004a). *Niger Delta Regional Master Plan : Environment and Hydrology.* NDDC.

NDRMP. (2004b). *Niger Delta Regional Master Plan: Chapter 1. Niger Delta Region, Land and People.* Abuja: NDDC. Retrieved 2015 йил 28-May from http://www.nddc.gov.ng/masterplan.html

NDRMP. (2004c). *Niger Delta Regional Master Plan. Chapter 2: Regional developement efforts.* Abuja: NDDC. Retrieved May 28, 2015, from http://www.nddc.gov.ng/masterplan.html

NDRMP. (2004d). *Niger Delta Regional Master Plan: Environment and Hydrology Report.* Port Harcourt, Nigeria: NNDC.

NEMA. (2010a). *National Disaster Management Framework: Thematic Area 1.* Retrieved 01 14, 2013, from http://www.nema.gov.ng/documentation/ndmf/forward/acroabr/thematic-area-1.aspx

NEMA. (2010b). *National disaster management frame work: thematic area 1. Institutional capacity for disaster management.* Retrieved 01 14, 2013, from http://www.nema.gov.ng/documentation/ndmf/forward/acroabr/thematic-area-1.aspx

NEMA. (2010c). Retrieved from ://www.nema.gov.ng/documentation/ndmf/forward /acroabr/thematic-area-1.aspx

NEMA,Nigeria, N. (2013). *NIGERIA: Post-Disaster Needs Assessment 2012 Floods.* Abuja: World Bank. Retrieved August 09, 2016, from https://www.gfdrr.org/sites/gfdrr/files/NIGERIA_PDNA_PRINT_05_29_2013_WEB.pdf

NEST. (2011). *Reports of Research Projects on Impacts and Adaptation.* Ibadan, Nigeria: Building Nigeria's Response to Climate Change (BNRCC). Retrieved July 26, 2015

NEST, & Woodley, E. (2011). *Reports of Pilot Projects in Community-based Adaptation - Climate Change in Nigeria. Building Nigeria's Response to Climate Change (BNRCC).* Ibandan, Nigeria: BNRCC. Retrieved from ISBN 978-0-9877568-9-3

Niang, I., Ruppel, O., Abdrabo, M., Essel, A., Lennard, C., Padgham, J., & Urquhart, P. (2014). Part B: Regional Aspects. Contribution of Working GroupII to the Fifth Assessment Report of the

Intergovernmental Panel on Climate Change. In V. Barros, C. Field, D. Dokken, M. Mastrandrea, K. Mach, T. Bilir, . . . L. White (Eds.), *Climate Change 2014: Impacts, Adaptation, and Vulnerability.* (pp. 1199-1265). Cambridge, United Kingdom and New York, NY, USA: IPCC. Retrieved 08 12, 2016, from http://www.ipcc.ch/pdf/assessment-report/ar5/wg2/WGIIAR5-Chap22_FINAL.pdf

Nicholls, R., Wong, P., Burkett, V., & Codi, J. (2007). *Coastalsystems and low-lying areas. Climate Change 2007: Impacts, Adaptation and Vulnerability. Contribution of Working Group II to the Fourth Assessment Report of the Intergovernmental Panel on Climate Change.* CambridgE, UK: Cambridge university press.

Nicolls, R., & Mimura, N. (1998). Regional issues raised by sea-level rise and their policy implications. *Journal of Climate Change,* 5-18.

Nicolls, R., Hoozemans, F., & Marchand, M. (1999). Increasing #ood risk and wetland losses due to global sea-level rise: regional and global analyses. *Global and Environmental Change, 9,* S69-S87. doi:10.1016/S0959-3780(99)00019-9

NOAA. (2012). *Detailed methodology for mapping sea level rise inundation.* Charleston, South Carolina 2: NOAA-Coastal services centre. Retrieved November 18, 2015, from https://coast.noaa.gov/digitalcoast/_/pdf/ElevationMappingConfidence.pdf

NOAA. (2016). *Causes of coastal flooding.* Retrieved December 25, 2016, from Storm surge and coastal inundation: http://www.stormsurge.noaa.gov/overview_causes.html

NOAA, S. I. (2015). *Jason3.* Retrieved May 29, 2015, from http://www.nesdis.noaa.gov/jason-3/?CFID=731ecb89-8379-48fc-ad50-959546e71739&CFTOKEN=0

NPC. (2010). *CENSUS ,Population distribution by state, sex, LGA and senatorial district. 2006 priority tables.* Retrieved 2013 йил 14-01 from Available at: http://www.population.gov.ng

Nwilo, C. (1997). Managing the impact of Storm Surges in Victoria Island Nigeria. *Destructive Waters: water caused natural disasters, their abatement and control* (pp. 325-330). Califonia: IAHS.

Ogba, C., & Utang, P. (2007). vulnerability and adaptation of nigeria's niger delta coast settlements to climate change induced sea level rise. *Strategic Integration of Surveying Services FIG Working week 2007.* Hong Kong SAR, China: FIG. Retrieved from https://www.fig.net/resources/proceedings/fig_proceedings/fig2007/papers/ts_7b/ts07b_06_o gba_utang_1342.pdf

Ogoro, M. (2014). Spatio – Temporal Changes in the Geomorphic Shoreline of Bonny Island. *Journal of Research in Humanities and Social Science, 2*(11), 75-80.

Olomoda, I. (2012). Geostrategic Plan for Mitigation of Flood Disaster in Nigeria. *Special Publication of the Nigerian Association of Hydrological Sciences, 2012*. Retrieved from http://journal.unaab.edu.ng/index.php/NAHS

Oosthoek, K. (2006). *Dutch river defences in historical perspective.* Retrieved February 23, 2017, from Environmental history resources: https://www.eh-resources.org/dutch-river-defences-in-historical-perspective/

Orupabo, S. (2008, January 01). *Coastline Migration in Nigeria.* Retrieved January 12, 2017, from Hydro International. Surveying in all waters: https://www.hydro-international.com/content/article/coastline-migration-in-nigeria

Owe, M., Brubaker, K., Ritchie, J., & Albert, R. (2001). *Remote sensing and Hydrology, 2000.* IAHS.

Ozyurt, G., & Ergin, A. (2009). Application of sea level rise Vulnerability Assessment Model to Selected Coastal Areas of Turkey. *Journal of Coastal Research*(56), 248-251.

Ozyurt, G., & Ergin, A. (2010).) Improving Coastal Vulnerability Assessments to Sea-Level Rise: A New Indicator Based Methodology for Decision Makers. *JOURNAL OF COASTAL RESEARCH, 56*, 248-251.

Papa, F., Bala, S. K., Pandey, R. K., Durand, F., Gopalakrishna, V. V., Rahman, A., & Rossow, W. B. (2012). Ganga-Brahmaputra river discharge from Jason-2 radar altimetry: An update to the long-term satellite-derived estimates of continental freshwater forcing flux into the Bay of Bengal. *Journal of Geophysical Research, 117*(C11021). doi:10.1029/2012JC008158

Pavelsky, T., Morrow, R., Peterson, C., Andral, A., Bronner, E., & Srinivasan, M. (2015). *SWOT 101: A Quatum Improvement of Oceanography and Hydrology from the Next Generation Altimeter Mission.* Retrieved May 29, 2015, from Surface Water and Ocean Topography: https://swot.jpl.nasa.gov/files/swot/SWOT-101_Jan2015.pdf

Pendelton, E., Barras, J., Williams, S., & Twitchell, D. (2010). *Coastal Vulnerability Assessment of the Northern Gulf of Mexico to Sea-Level Rise and Coastal Change.* USGS. Retrieved from HTTP://pubs.usgs.gov/of/2010/1146

Penton, D. J., & Overton, I. C. (2007). Spatial modelling of floodplain inundation combining satellite imagery and elevation models. *MODSIM 2007 International Congress on Modelling and Simulation.* Modelling and Simulation Society of Australia and New Zealand.

Pereira-Cardenal, S. J., Riegels, N. D., Berry, P. A., Smith, R. G., Yakovlev, A., Siegfried, T. U., & Bauer-Gottwein, P. (2011). Real-time remote sensing driven river basin modeling using radar altimetry. *Hydrology and Earth sciences*, 241-254. doi:10.5194/hess-15-241-2011

Pethick, J., & Orford, J. (2013). Rapid rise in effective sea-level in southwest Bangladesh: Its causes and contemporary rates. *Global and Planetary Change, 111*(2013), 237–245. doi:dx.doi.org/10.1016/j.gloplacha.2013.09.019

Pfeffer, W. T., Harper, J. T., & O'neel, S. (2008, September 5). Kinematic Constraints on Glacier Contributions to 21st-Century Sea-Level Rise. *Science, 32*(5894), 1340-1343 . doi:doi: 10.1126/science.1159099

Popescu, I., Cioaca, E., Pan, Q., Jonoskia, A., & Hanganu, J. (2015). Use of hydrodynamic models for the management of the Danube Delta wetlands: The case study of Sontea-Fortuna ecosystem. *Environmental Science & Policy, 46*, 48–56. doi:dx.doi.org/10.1016/j.envsci.2014.01.012

Price, R. (2009). An optimized routing model for flood forecasting. *water Resources Research, 45*, W02426. doi:10.1029/2008WR007103

Quinn, P., Hewett, C.J.M., Muste, M., & Popescu, I. (2010). Towards new types of water-centric collaboration. Proceedings of the Institution of Civil Engineers. *Water Management, 163*(1), 39-51. doi:DOI: 10.1680/wama.2010.163.1.39

Rahmstorf, S. (2007, January 19). A Semi-Empirical Approach to Projecting Future Sea-Level Rise. *Science, 315*(5810), 368-370 . doi:DOI: 10.1126/science.1135456

Ranasinghe, R., Duong,, T., Uhlenbrook, S., Roelvink, D., & Stive, M. (2012). Climate-change impact assessment for inlet-interrupted coastlines. *Nature Climate Change, 3*, 83–87. doi:10.1038/nclimate1664

Ričko, M., Birkett, ,. C., Carton, ,. J., & Crétauxc, J.-F. (2012). Intercomparison and validation of continental water. *Journal of Applied Remote Sensing, 6*. Retrieved from spiedigitallibrary.org/ on 03/12/2013 Terms of Use: http://spiedl.org/terms

Rijsberman, F. (1996). Rapportteurs statement: coastal resources. In S. S. Media, *Adapating to climate change: an international perspective* (pp. 279-282). New York, USA: Springer. doi:10.1007/978-1-4613-8473-1

Rosmorduc, V. (2012). EO Information Services in support of West Africa Coastal vulnerability. Service 2 : Sea Level Height & currents. Washington DC: World bank. Retrieved from http://siteresources.worldbank.org/EXTEOFD/Resources/8426770-1335964503411/SUR-C-West_Africa_Coastal_vulnerability_Service2.pdf

Santos da Silva, J., Roux, E., Filho, O., Bonne, M. P., Seyler, F., & Calmant, S. (2007). 3D selection of Envisat data for improved water stage times series on the Rio Negro and adjacent Wetlands (Amazon Basin). *2nd Hydrospace Workshop*. Geneva: ESA.

Santos da Silver, J., Calmant, S., Seyler, F., Lee, H., & Shum, C. (2012). Mapping of the extreme stage variations using ENVISAT altimetry in the Amazon basin rivers. *International Water Technology Journal, IWTJ, 2*(1), 14-25. Retrieved from http://iwtj.info/wp-content/uploads/2012/11/V2-N1-p2.pdf

Sanyal, J., & Carbonneau, P. (2012). Low-cost, open access flood inundation modelling with sparse data: A case study of the Lower Damodar River Basin, India. EGU.

Sarhadi, A., Soltani, S., & Modarres, R. (2012). Probabilistic flood inundation mapping of ungauged rivers: Linking GIS techniques and frequency analysis. *Journal of Hydrology, 458-459*, 68-86. doi:10.1016/j.jhydrol.2012.06.039

Schielen, R. M. (2010). Flood management. In R. C. Ferrier, & A. Jenkins (Eds.), *A handbook of catchment management* (pp. 51 - 76). Wiley-Blackwell.

Schumann, G. J.-P., Bates, P., Neal, J., & Andreadis, K. (2015). Measuring and Mapping Flood processes. In P. Paron, & G. Di Baldassare (Eds.), *Hydro-Meteorological Hazards, Risks, and Disasters* (pp. 35-64). Amsterdam,Netherlands: Elsevier.

Schumann, G. J.-P., Neal, J., Voisin, N., Andreadis, K., Pappenberger, F., Phanthuwongpakdee, N., . . . Bates, P. (2013). A first large-scale flood inundation forecasting model. *Water Resources Research, 49*, 6248–6257. doi:10.1002/wrcr.20521, 2013

Schumann, G., Matgen, P., Cutler, M., Black, A., Hoffmann, L., & Pfister, L. (2008). Comparison of remotely sensed water stages from LiDAR,topographic contours and SRTM. *Photogrammetry and Remote Sensing, 63*, 283–296. doi:10.1016/j.isprsjprs.2007.09.004

Schumann, G., Matgen, P., Hoffmann, L., Hostache, R., Pappenberger, F., & Pfister, L. (2007). Deriving distributed roughness values from satellite radar data for flood inundation modelling. *Journal of Hydrology, 344*(1-2), 96–111. doi:10.1016/j.jhydrol.2007.06.024

Schumann, G., Matgen, P., Pappenberger, F., Black, A., Cutler, M., Hoffmann, L., & Pfister, L. (2006). The REFIX model: remote sensing based flood modelling. *ISPRS Commission VII Mid-term Symposium "Remote Sensing: From Pixels to Processes.* Enschede, the Netherlands: ISPRS commission.

Seyler, F., Calmant, S., Santos da Silva, J., León, G. J., Frappart, F., Bonnet, M.-P., . . . Seyler, P. (2009). New perspectives in monitoring water resources in large tropical transboundary basins based on the combined used of remote sensing and radar altimetry. *Improving Integrated Surface and Groundwater Resources Management in a Vulnerable and Changing World*, pp. 282-288. doi:10.13140/2.1.5101.0569

Shell, E. (2004). *Environmental impact assessment for the utorogu NAG*. Shell Nigeria. Retrieved 2015 йил 26-May from http://s06.static-shell.com/content/dam/shell/static/nga/downloads/environment-society/eia-reports/utorogu-nag-wells-eiareport.pdf

Shell, E. (2006). *Environmental Impact Assessment of Rumuekpe (Oml 22) and Etelebou (Oml 28) 3d Seismic Survey*. The Shell Petroleum Development Company of Nigeria Limited. From http://s04.static-shell.com/content/dam/shell-new/local/country/nga/downloads /pdf/rumuekpe-eia-report.pdf

Shell, N. (2015). *Environment and society:Oil spill data*. Retrieved May 22, 2015, from http://www.shell.com.ng/environment-society/environment-tpkg/oil-spills.html

Siddique-E-Akbor, A. H., Hossain, F., Lee, H., & Shum, C. K. (2011). Inter-comparison study of water level estimates derived from. *Remote Sensing of Environment, 115*(6), 522–1531. doi:10.1016/j.rse.2011.02.011

Skakun, S. (2010). A neural network approach to flood mapping using satellite imagery. *Computing and Informatics, 29*, 1013–1024.

Smith, L. C. (1997). Satellite Remote Sensing of River Innundation Area, Stage and Discharge: A review. *Hydrological Processes, 11*, 1427-1439.

Smith, L., Liang, Q., James, P., & Lin, W. (2015). Assessing the utility of social media as a data source for flood risk management using a real-time modelling framework: Assessing the utility of social media for flood risk management. *Journal of Flood Risk Management*, 1-11. doi:10.1111/jfr3.12154

Sorensen, R., Weisman, R., & Lennon, G. (1984). Control of Erosion, Inundation and Salinity Intrusion Caused by Sea Level Rise. In BARTH, M., & TITUS, J. (Eds.), *Green House Effect and Sea Level Rise: A Challenge for This Generation. 1st ed* (pp. 179-214). Van Nostrand Reinhold Company.

Stephens, E., Bates, P., Freer, J., & Mason, D. (2012). The impact of uncertainty in satellite data on the assessment of flood inundation models. *Journal of Hydrology, 414-415*, 162-173. doi:10.1016/j.jhydrol.2011.10.040

Structurae. (1997). *Maeslant Barrier*. Retrieved from https://structurae.net/about/terms/

Sugiura, A., Fujioka, S., Nabesaka, S., Sayama, T., Iwami, Y., Fukami, K., . . . Takeuchi, K. (2013). Challenges on modelling a large river basin with scarce data: A case study of the Indus upper catchment. *20th International Congress on Modelling and Simulation, Adelaide, Australia, 1–6 December 2013* . Adelaide, Australia. Retrieved from www.mssanz.org.au/modsim2013

Sun, W., Ishidaira, H., & BasTola, S. (2009). Estimating discharge by calibrating hydrological model against water surface width measured from satellites in large ungauged basins. *Annual Journal of Hydraulic Engineering, 53*, 49-54.

Sun, W., Ishidaira, H., & BasTola, S. (2010). Towards improving river discharge estimation in ungauged basins: calibration of rainfall-runoff models based on satellite observations of river flow width at basin outlet. *Hydrology and Earth Sysytem Sciences, 14*, 2011–2022. doi:10.5194/hess-14-2011-2010

Syvitski, J. P. (2008). Deltas at risk. *Sustainable Science, 3*, 23-32. doi:10.1007/s11625-008-0043-3

Syvitski, J. P., Overeem, I., Brakenridge, R. G., & Hannon, M. (2012). Floods, floodplains, delta plains — A satellite imaging approach. *Sedimentary Geology, 267-268*, 1-14. doi:10.1016/j.sedgeo.2012.05.014

Tarpanelli, A., Barbetta, S., Brocca, L., & Moramarco, T. (2013). River Discharge Estimation by Using Altimetry Data and Simplified Flood Routing Modeling. *Remote Sensing, 5*, 4145-4162. doi:10.3390/rs5094145

Thieler, E., & Hammer-Kloss, E. (1999). *National Assessment of Coastal Vulnerability to Future Sea-Level Rise: Preliminary Results for the U.S. Atlantic Coast.* US Geological Survey.

Tides4Fishing. (2016). *Tides and solunar charts.* Retrieved march 01, 2016, from http://www.tides4fishing.com/af/nigeria/ford-point

Tilmans, W., Jakobsen, P., & LeClerc, J.-P. (1991). *Coastal erosion in the Bight of Benin: a critical review.* Delft: Delft Hydraulics.

Tol, S. (2007). the double trade off between adaptation and mitigation for sea level rise: an application of fund . mitigation and adaptation strategies. *Global Change, 12*, 741-753.

Tulloch, J. (2016). *Dutch flood protection: Taming the water wolf.* Retrieved February 24, 2017, from Allianz: https://www.allianz.com/en/about_us/open-knowledge/topics/environment/articles/100311-dutch-flood-protection-taming-the-water-wolf.html/

Twmasi, Y., & Merem, E. (2006). GIS and Remote Sensing Applications in the Assessment of Change within a Coastal Environment in the Niger Delta Region of Nigeria. *International Journal of Environmental Research and Public Health, 3*(3), 98-106.

UK Hydrographic office: Almiralty Easy Tide. (n.d.). *Your EasyTide Prediction .* Retrieved 01 29, 2016, from http://www.ukho.gov.uk/easytide/easytide/ShowPrediction.aspx?PortID=3663&PredictionLength=7

UNFPA. (2008). *Population and Climate Change Framework of UNFPA's Agenda.* Retrieved 03 05, 2013, from Climate change: http://www.unfpa.org/pds/climate_change_unfpa.pdf

US Army Corps of Engineers. (2009). *Louisiana coastal protection and restoration (LACPR) final technical report.* Mississippi Valley U.S.A: U. S. Army Corps of Engineers. Retrieved from http://www.mvn.usace.army.mil/Portals/56/docs/environmental/LaCPR/LACPRFinalTechnic alReportJune2009.pdf

Utretcht University: Faculty of Geosciences. (2007). *Rhine - Muese delta studies: Flooding.* Retrieved March 06, 2017, from http://www.geo.uu.nl/fg/palaeogeography/results/flooding

Uy, N., Takeuchia , Y., & Shawa, R. (2011). Local adaptation for livelihood resilience in Albay, Philippines. *Environmental Hazards, 10*(2). doi:10.1080/17477891.2011.579338

Uyigue, E. (2007). *Climate change in the Niger delta.* Retrieved from CIEL.ORG: www.ciel.org/Publications/Climate/CaseStudy_Nigeria_Dec07.pdf

Uyigue, E. (2009). *The Changing Climate and the Niger Delta.* CREDC. Retrieved 2015 йил 20-11 from http://www.climatefrontlines.org/sites/default/files/The%20Changing%20Climate%20 and%20the%20Niger%20Delta.pdf

Uyigue, E., & Agwo, M. (2007). *Coping with climate change and environmental degradation in the Niger delta of southern Nigeria.* Benin city, Nigeria: CREDC.

Uzukwu, P., Leton, P., & Jamabo, N. (2014). Survey of the physical charecteristics of the upper reach of the New-Calabar river, Niger delta Nigeria. *Trends in Aplied Sciences Research, 9*(9), 494-502. doi:10.3923/tasr.2014.494.502

Van Bentum, K. (2012). *The Lagos coast - Investigation of the long-term morphological impact of the Eko Atlantic City project.* T.U Delft. Delft: Delft University of Technology. Retrieved from http://resolver.tuDelft.nl/uuid:794318f9-9279-4a3c-9030-2f9fab7c6562

van der Burgh, L. (2008). *Risk and coastal zone policy: example from the Netherlands. .* Retrieved March 07, 2017, from http://www.coastalwiki.org/wiki/Risk_and_coastal_zone_policy:_example_from_the_Netherl ands

Van Heerden, I. (2007). The Failure of the New Orleans Levee System Following Hurricane Katrina and the Pathway Forward. *Public Administration Review, 67*(2007), 24–35. Retrieved from www.jstor.org/stable/4624679

Van, P., Popescu, I., Van Griensven, A., Solomatine, D., Trung, N., & Green, A. (2012). A study of the climate change impacts on fluvial flood propagation in the Vietnamese Mekong Delta. *Hydrology and Earth Systems Science, 16*, 4637-4649. doi:doi:10.5194/hess-16-4637-2012

Vermeulen, C. J., Barneveld, H. J., Huizinga, H. J., & Havinga, F. J. (2005). Data-assimilation in flood forecasting using time series and satellite data. *International conference on innovation advances and implementation of flood forcasting technology*. Tromso: ACTIF/Floodman/FloodRelief.

Villladsen, H., Andersen, O., & Stenseng, L. (2014). Annual cycle in lakes and rivers from Cryosat-2 altimetry - the Brahmaputra river. *International Geoscience and Remote Sensing Symposium (IGARSS)* (pp. 894-897). Quebec, canada: IEEE. doi:978-1-4799-5775-0/14

Wesselink, A., Abu Syed, M., Chan, F., Duc Tran, D., Huq, H., Huthoff, F., . . . Zegwaard, A. (2015). *International Journal of Water Governance, 3*(4), 25-46. doi:10.7564/15-IJWG90

Westahoff, R., Huizinga, J., Kleuskens, M., Burren, R., & Casey, S. (2010). ESA Living Planet Symposium. *686*. Bergen, Norway: ESA Communications.

Williams, S., & Ismail , N. (2015). Climate Change, Coastal Vulnerability and the Need for Adaptation Alternatives: Planning and Design Examples from Egypt and the USA. *Marine Science and ENgineering, 3*(2015), 591-606. doi:10.3390/jmse3030591

Woldemicheal, A., Degu, A., Siddique-E-Akbor, A., & Hossain, F. (2010). Role of Land–Water Classification and Manning's Roughness Parameter in Space-Borne Estimation of Discharge for Braided Rivers: A Case Study of the Brahmaputra River in Bangladesh. *Ieee journal of selected topics in applied earth observations and remote sensing*. Doi:10.1109/jstars.2010.2050579

Woody, T. (2015). *Will the 'Great Wall' of New Orleans Save It From the Next Killer Hurricane?* Retrieved January 22, 2017, from http://www.takepart.com/feature/2015/08/17/katrina-new-orleans-walled-city

World Bank. (2016). *inception report: Cost of Coastal Environmental Degradation, Multi Hazard Risk Assessment and Cost Benefit Analysis*. Belgium: World bank.

World Population Review. (2017, February 08). *Lagos Population 2017*. Retrieved 02 19, 2017, from World Population Review: http://worldpopulationreview.com/world-cities/lagos-population/

Yan, K., Di Baldassarre, G., Solomatine, D., & Schumann, G. (2015). A review of low-cost space-borne data for flood modelling:topography, flood extent and water level. *Hydrological processes, 29*(14), 3368-3387. doi:10.1002/hyp.10449

Yan, K., Di Baldassarre, G., Solomatine, D., & Schumann, G. (2015). A review of low-cost space-borne data for flood modelling:topography, flood extent and water level. *Hydrological processes*. doi:10.1002/hyp.10449

Yang, C., & Ouchic, K. (2012). Analysis of bar morphology using multi-temporal and multi-sensor satellite images: Example from the Han Estuary, Korea. *Marine Geology, 311-314*, 17–31. doi:10.1016/j.margeo.2012.04.004

Yin, J., Yin, Z., Wang, J., & Xu, S. (2012). National Assessment of Coastal Vulnerability to Sea-Level Rise for the Chinese coast. *Journal of coastal Conservation, 16*, 123-133.

Zhang, J., Xu, K., Yang, Y., Qi, L., Hayashi, S., & Watanabe, M. (2006). Measuring water storage fluctuations in Lake Dongting, China, by Topex/Poseidon satellite altimetry. *Environmental Monitoring and Assessment, 115*, 23-37. doi:10.1007/s10661-006-5233-9

Acknowledgements

First and foremost my gratitude and appreciation go to Professor Arthur Mynett, my promotor, challenger and supporter. This PhD would never have been possible without your help and leadership. Thank you for the pertinent questions that always reminded me about the basics. Your inputs kept me on track, and made me think not just as an engineer but also as a scientist.

I am grateful to my supervisor and teacher Dr Ioana Popescu who always had time to listen to me, read, and edit my work. I learnt a lot from you about hydrodynamics, the engineering of channel flows and their interactions with the environment. Thank you for giving me opportunities to improve my knowledge of modelling and data analysis. Your invaluable advice and corrections made my PhD easier.

I thank the Nigerian National Space Research and Development Agency (NASRDA) for the financial support to start my PhD. In 2010 the federal government of Nigeria asked NASRDA to use satellite technology to study 'Impacts of sea level rise for the Niger delta' and suggest adaptation measures, which made NASRDA allocate money to start this study.

I thank Professor Dimitry Solomatine for sharing great scientific insights on uncertainty analysis during my MSc. Thank you for the many jokes that helped me laugh off my worries as I worked on my PhD research. Together with Dr Gerald Corso, Dr Leonardo Segura and his beautiful wife, I thank you all for the unforgettable support when I fell ill during the HIC conference trip to South Korea. You all saved my life, and I can never thank you enough.

To my IHE Delft family, I say a big thank you. When I arrived Delft in 2011 I had no idea I would end up with such a large group of friends. In IHE, I met people from all works of life who were as curious and ready to learn as I was. I thank all colleagues in the hydro-informatics core group that have been there every day during these four years. My special thanks to Mario, Oscar, Quan, Nieler and Yared, for the laughs and small talk that kept me smiling through everything.

I thank my IHE friends who made life so much better for me in IHE: Marmar, Hassana, Pedi, Sondos, Mawati, Jakia, Maria, Reem, Sara, Emaan, and Mohan. I thank my friends and confidants Alifta and Khafila for all the emotional support and many happy moments in the

Netherlands. Many thanks to IHE Delft staff who made life better for me in their own unique ways; notably, Peter Heerings, Jolanda Boots, Jos Bult, and Marrielle van Erven.

I am grateful to colleagues and friends at NASRDA who helped with data and words of encouragement. I particularly thank John Nwagwu, Sadiya Baba and Mustapha Aliyu. My heartfelt gratitude also goes to colleagues at the National Centre for Remote Sensing (NCRS) Jos Nigeria, for NigeriaSat 2, and NigeriaSatX data - my work wouldn't have been possible without your help.

My eternal gratitude goes to my family who had to sacrifice so much to enable me do the PhD. I thank my daughters Hadiza and Hassana, who are my support group and personal cheer leaders. Many thanks to my husband Hamza Makarfi, my friend and pillar who made so many changes and pushed so many limits to support my PhD. I thank my parents Mr and Mrs G.J Sallah, for their love and constant encouragement. Many thanks to my sister Mabi Sallah Lohor and my brother Boolnaan Sallah.

My PhD journey followed a narrow alley whose course had been foreseen by great people I was fortunate to meet in life. Professor Funso Olorunfemi, Professor S.O Fadare, and Professor S.O Mohammed - this PhD is a realisation of your vision for me, thank you for believing in me even before I could believe in myself. To you all I owe much gratitude, and no words can adequately explain my indebtedness to your good hearts.

Much thanks to IHE Delft for giving me a chance to do my PhD, I am grateful to the management for their patience with me through the many challenges I faced.

Finally, I thank Allah for making it possible for me to undertake this study. Alhamdulillah.

About the author

Zahrah Naankwat Musa was born in Gboko, Nigeria. She started her education at a tender age when she followed school children walking past her home on their way to school. Since then she has kept her enthusiasm for knowledge and continues to learn new things at every opportunity. In 1998 she graduated with a B.Eng in electrical/electronic engineering from the Federal University of Agriculture Makurdi (UAM) Nigeria. After her compulsory youth service from 1998 – 1999, she served as a part-time Physics lecturer
and Physics teacher Federal College of Land Resources Technology and the College of Mary Immaculate both in Jos, Nigeria. In 2001 she started working as a computer analyst with the National Centre for Remote Sensing (NCRS) in Jos Nigeria under the Nigerian National space Research and Development Agency (NASRDA). To develop her professional knowledge she studied for a postgraduate diploma in Remote Sensing and GIS from the African Regional Centre for Space Technology Education in Ile-Ife Nigeria, in 2006, and an MSc in Space studies from the International Space University Strasbourg, France. In 2013, inspired by her research in hazard and environmental management, she studied for another MSc in Hydro-informatics from the IHE Delft institute for water education in Delft, the Netherlands.

After being accepted by her promotor at IHE Delft/TU Delft, Zahrah proceeded with her PhD research which is based on satellite data and combined GIS/Remote sensing, hydrodynamic modelling and analytical techniques to propose adaptation and mitigation options for impacts of sea level rise on ungauged/ poorly gauged deltas. The research work included analysis of methodologies used for flood/inundation modelling and mapping, assessment of sources of unconventional data that can be used to bridge data gaps, and assessment of both vulnerability and resilience of the study area to impacts of sea level rise. Zahrah is married to Hamza Musa Makarfi and has two daughters Hadiza and Hassana.

Netherlands Research School for the
Socio-Economic and Natural Sciences of the Environment

D I P L O M A

For specialised PhD training

The Netherlands Research School for the
Socio-Economic and Natural Sciences of the Environment
(SENSE) declares that

Zahrah Naankwat Musa

born on 06 September 1974 in Gboko, Nigeria

has successfully fulfilled all requirements of the
Educational Programme of SENSE.

Delft, 06 April 2018

the Chairman of the SENSE board

Prof. dr. Huub Rijnaarts

the SENSE Director of Education

Dr. Ad van Dommelen

The SENSE Research School has been accredited by the Royal Netherlands Academy of Arts and Sciences (KNAW)

KONINKLIJKE NEDERLANDSE
AKADEMIE VAN WETENSCHAPPEN

The SENSE Research School declares that Ms Zahrah Naankwat Musa has successfully fulfilled all requirements of the Educational PhD Programme of SENSE with a work load of 39.7EC, including the following activities:

<u>SENSE PhD Courses</u>

o   Environmental research in context (2013)
o   Research in context activity: 'Initiating and organizing seminar on "Use of satellite data and information to support living with sea level rise on the subsiding Niger delta", Abuja, Nigeria' (2014)
o   Where there is little data: How to estimate design variables in poorly gauged basins (2016)

<u>External training at a foreign research institute</u>

o   PhD Summer School, Microsoft research, Cambridge, United Kingdom (2016)

<u>Management and Didactic Skills Training</u>

o   Student organiser IAHR conference, The Hague, The Netherlands (2015)
o   Teacher and teaching assistant MSc. course 'Catchment modelling, use of Hecras for river modelling' (2015)
o   Supervising two MSc students with thesis entitled 'Use of synthetic cross-sections for river flow modelling in data scarce areas: the case study of Niger River' (2016) and 'Effect of flood adaptation measures in mitigating flooding problems in the Niger rivers, upstream Onitsha Bridge' (2016)
o   Member of the Unesco-ihe gender and ethics committee (2016)

<u>Selection of oral Presentations</u>

o   *How vulnerable is the Niger delta to inundation from sea level rise*. Geospatial world forum, 05-09 May 2014 Geneva, Switzerland
o   *Uncertainty in hydrodynamic modeling of flooding in the lower Niger river*. Simhydro conference, 11-13 June 2014, Nice, France
o   *Modeling the effect of sea level rise on flooding in the lower Niger river with downstream sea level rise*. Hydroinformatics conference, 18-21 August 2014, New York, USA
o   *Approach on modeling complex deltas in data scarce areas*. Hydroinformatics conference, 22-26 August 2016, Incheon, South Korea

SENSE Coordinator PhD Education

Dr. Peter Vermeulen